The Structure of the Planets

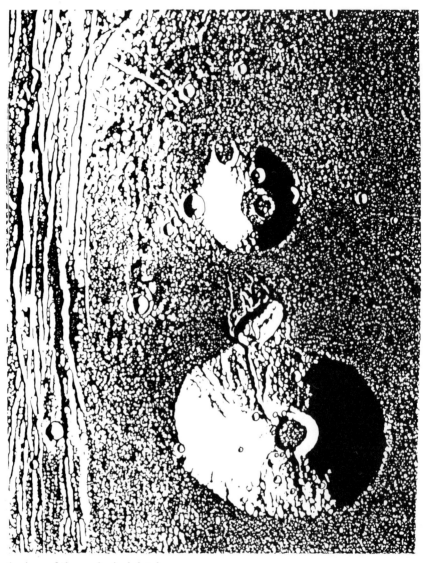

A piece of the geological database.

View of a 300 × 400 km portion of Mars in the north Tharsis region near 100 W, 25 N, showing a lava flood plain, strongly crinkled to the west (Ceraunius Fossae), with two prominent shield volcanoes (Ceraunius, lower, and Uranius, upper) and several impact craters.

(Adapted, with a little artist's licence, from a photo-mosaic taken by Viking satellite, 1975; reference JPL 211-5639, National Space Science Data Centre, Greenbelt, Maryland, USA, with permission.)

The Structure of the Planets

JOHN ELDER
Professor Emeritus
Geology Department
University of Manchester

1987

ACADEMIC PRESS
Harcourt Brace Jovanovich, Publishers
London   Orlando   San Diego   New York
Austin   Boston   Sydney   Tokyo   Toronto

ACADEMIC PRESS INC. (LONDON) LTD
24/28 Oval Road, London NW1 7DX

United States Edition Published by
ACADEMIC PRESS INC.
Orlando, Florida 32887

Copyright © 1987 by
ACADEMIC PRESS INC. (LONDON) LTD.

All rights reserved. No part of this book may be reproduced
in any form by photostat, microfilm, or any other means,
without written permission from the publishers

British Library Cataloguing in Publication Data

Elder, John, 1933–
 The structure of the planets.—(Academic
Press geology series)
 1. Planets
 I. Title
 523.4   QB601

ISBN 0-12-236452-X

Typeset and printed by
Galliard (Printers) Ltd, Great Yarmouth

PREFACE

This monograph is for scientists interested in the nature of our planet as seen in the perspective of the planetary bodies of our solar system. It draws together work from astronomy and geology.

This book could not have been written 20 years ago. Since that time our knowledge of the Earth itself and of its place in the family of the solar system has made several steps forward. The Earth is now seen as a vigorous and mobile body. Furthermore, the earlier pages of the geological record are opening up and they reveal a new picture of the evolution of our planet. Detailed studies of the planets and their environment and in some cases of their surfaces, in this post-sputnik era of space exploration, have shown us a vista of the planets and the solar system not dreamed of before. Notable advances have also been made in our knowledge of the origin and evolution of stars, and although the merest cartoon of the origin of the solar system is yet possible at least we can be confident about some of its features. In support of the study of these objects of interest have been substantial developments in technical subjects, e.g. computation and geophysical fluid dynamics.

The foundation of all these studies goes back a long way and it is of interest to note how often one refers to the work of the 18th century (orbits, figure of the Earth etc.). Much of this early work, however, had no flesh to put on its bones—the ideas were often in an advanced state but there were no data. The past 20 years have put a lot of flesh on the bones.

My own interest in this subject goes back to 1946 when, in Timaru, New

Zealand, as a schoolboy in the third form I had the luck to attend a WEA course of ten lectures on astronomy by Albert Jones (already then one of the world's outstanding amateur astronomers). Under his tutelage I became a regular observing member of the RNZAS Variable Star Section, so that I count among my friends L2 Puppis, R and S Carina, R Doradus, S Pavonis, T Centauri, Mira etc.; I have seen numerous comets, all the planets, multitudes of sunspots etc.

I lectured on this subject at Manchester University from 1970 to 1984 but the incentive to write this book came initially from the need to prepare a course on planetary geology for third year geology and physics students.

This book is, however, more of a scientific essay with the dominant theme of the role of phase changes in controlling the development of planetary structure. The focus of interest is in the nature and structure of planetary bodies, and in particular those physical and chemical processes which have produced the individual characteristics of the planets. A feature of the work is the description of all the planetary bodies from a uniform point of view—and particular emphasis is placed on the constraints on our ideas which arise from measurements made external to the planets and from our knowledge of their geological development.

The approach emphasizes the material, what it is and how it got there. The viewpoint is of a system continually rearranging itself as it moves through its elaborate phase space, exploiting and using up its many degrees of freedom. This is a picture of continually changing structures—conceptually like a movie film rather than a museum full of sculptures. Much of the picture is speculative but there are some moments when the system can be identified by measurement. The keystone of the work is the calibration of the fractionation model of the chemical development of the terrestrial planets with data from upper mantle rocks.

Two key models form the framework of this study. I refer to them as the "disc" model and the "pool" model. The disc model of Chapter 3 provides the setting for the discussion of the emergence of the planets and their initial structures during the hydrogen era. The pool model of Chapter 6 focuses attention on the development of the structures of the terrestrial planets through their chemical and structural phase space during the geological era.

The style of presentation is quantitative but the emphasis is on the objects of interest, the planets, rather than on details of the physics, chemistry or mathematics—nevertheless I am interested in showing how and with what certainty our knowledge of planetary structure is obtained. Actual measured data is the basis of the work. Nevertheless the reader should regard the details of the numerical models with extreme scepticism.

There is a voluminous literature related to the subject of this book. I have decided to refer explicitly only to those matters which bear directly on my theme. For students who are just beginning to take an interest in these matters there are several good texts available to provide background information. For the expert it would be tedious to have to put up with yet another regurgitation of familiar ideas.

I wish to acknowledge with thanks the assistance of Barry Ashcroft (postgraduate student, Chemistry Department, University of Auckland, New Zealand), Dr Graham Fraser (reader, Physics Department, University of Canterbury, New Zealand), Dr Michael Henderson (senior lecturer, Geology Department, University of Manchester, England) and Prof. Jack Zussman (Geology Department, University of Manchester, England) for reading and commenting on the manuscript; of Barbara Guy for typing the manuscript and Alison Grey for drafting the figures. I am especially grateful for the geological education provided to me over many years by colleagues and students in the Geology Department, University of Manchester—I hope some of it shows.

J. W. Elder

CONTENTS

Preface v

Notation xv

1. Prologue

    I. Introduction 1
    II. An Isolated Gravitational System 1
    III. Identification of Gaseous and Stony Bodies 2
    IV. On the Rotundity of Large Bodies 4
    V. System Budget 5
        A. Atomic Species 6
        B. Mechanical Resources 6
        C. Energy Budget 7
    VI. Overview of Ideas About the Origin of the Solar System and the Planets 9
        A. Catastrophe 9
        B. Nebula Condensation 9
    VII. Viewpoint of this Book 12
    VIII. Outline of the Book 15

PART I. THE HYDROGEN ERA 17

2. Gestation of the Planets

I.	Introduction	19
II.	The Prelude	20
III.	A Turbulent Sheet	21
IV.	Onset of the Proto-Solar System	22
V.	Emergence of the Disc	26
VI.	The Role of Angular Momentum	26
VII.	Emergence of the Proto-Planets	27
VIII.	Sub-Planetary Material	30
	A. Particle Mass Estimates	31
	B. Collector Model	33
	C. Identification	34
	D. Obliteration of the Early Lunar Surface	36

3. Evolution of the Hydrogen Bodies

I.	Introduction	37
II.	Evolution of the Solar Disc	37
	A. Energy Considerations: Link to Time	38
	B. Disc Model	39
	C. Disc Development	43
III.	Evolution of the Jovian Planets	44
	A. Emergence of the Proto-Jovian Planets	45
	B. Development of Jupiter	47
	C. Physical Structure of the Jovian Planets	47

4. Parturition of the Proto-Terrestrial Planets

I.	Introduction	51
II.	Degassing	52
	A. General Considerations	53
	B. The Energy Budget	57
	C. Flushing the Proto-Terrestrial Discs	61
	D. Application to the Early Proto-Solar System	63
III.	Core Formation	68
	A. Emergence of the Proto-Terrestrial Central Bodies	70
	B. The Energy Budget	71
	C. Growth of the Cores of the Proto-Terrestrial Planets	72
IV.	Chronology	77

CONTENTS   xi

PART II. THE GEOLOGICAL ERA   79

5. Onset of Geological Time

I.	Introduction	81
II.	Pre-Crustal Fractionation	82
	The Surface Slag	84
III.	Simple Radiation Model of the Proto-Atmosphere–Crust	85
	The Net Transmissivity of the Proto-Geological Atmosphere	87
IV.	Amounts of H_2O, CO_2	88
	A. The Amount of H_2O in the Proto-Ocean	89
	B. The Amount of CO_2 in the Proto-Ocean	89
	C. The Amount of CO_2 in the Proto-Crust	89
	D. The H_2O, CO_2 Distribution	90
V.	The Proto-Crust	91
VI.	Model of the Early Atmosphere–Ocean–Crust for the Earth	92
	The Role of Hydrothermal Activity	93
VII.	Planetary Perspective	96
VIII.	Model for Venus	99
IX.	Model for Mars	100
	A. The Amounts of H_2O and CO_2 in the Proto-Atmosphere	100
	B. Model Sequence	100
	C. Allowance for Non-Uniform Surface Temperatures	101
	D. The CO_2 Sediments	102
	E. Properties of CO_2	102
	F. Mars Model Results	102
X.	Model for Titan	104

6. Chemical Development

I.	Introduction	107
II.	Fractionation Evidence	107
	A. The Primary Material	108
	B. Upper Mantle Fractionation Ratios	109
III.	The Mineral Assemblage	111
	Normative Mineral Model (CIPW Scheme)	111
IV.	Pool Model	113
	A. Model Behaviour	115
	B. Preliminary Application	116
V.	The Density Distribution	117
	A. Density of Rock Substance	118
	B. The Critical Level	119
	C. The Packing Factor	120
	D. The Mantle–Core Density Contrast	120

VI.	Simple Model with Uniform Packing	121
	A. The Central Chemically Homogeneous Mass	122
	B. On the Choice of f	123
VII.	The Inner Core	123
VIII.	Model Results for the Terrestrial Planets	125
IX.	The Density Distribution Sequence	131

7. Physical Development

I.	Introduction	133
II.	Earth Physical Structure Now	134
	A. The Equation of State	134
	B. Radial Variation of ρ_0	137
III.	The Model Relations	138
IV.	Method of Identification of the Models	140
V.	Identification	143
	A. Earth Identification	143
	B. Mars Identification	143
	C. Mercury Identification	144
	D. Venus Identification	144
	E. Moon Identification	145
VI.	Comparison of the Terrestrial Planets Now	146
VII.	Physical Structure Sequence	146
VIII.	The Jovian Moons	151
	A. The Properties of Ice	152
	B. The Small Ice–Rock Moons	153
	C. The Large Ice–Rock Moons Callisto, Ganymede and Titan	154
	D. The Large Stony Moons Europa and Io	155

8. Thermal History

I.	Introduction	157
II.	Model System	158
	Initial and Boundary Conditions	159
III.	Energy Contributions	159
	A. Contribution from Internal Thermal Energy	159
	B. Contribution from the Thermal Energy of the Core	160
	C. Gravitational Energy Available from Structural Change	160
	D. Energy Total	161
IV.	The Core Radius	162
V.	Temperature Distribution	162
	A. The Initial Global Mean Temperature θ_0	164
	B. The Upper Mantle Temperature Profile	164
	C. The Sublayer Base Temperature	165

CONTENTS

VI.	Partial Melting	165
	Role of Temperature Fluctuations	167
VII.	Heat Transfer	168
	A. Magma Convection	168
	B. "Solid" Convective Heat Transfer	169
	C. Contribution from Radioactivity	172
VIII.	Simple Model	173
	Simple Model Behaviour	174
IX.	Full Model	177
	Identification of Model Thermal Histories	178
X.	Geological Signature	188
	A. Energy Budget	188
	B. The Time Scales	189
	C. Remarks on Crustal Rearrangement	191
	D. Remarks on Volcanism	191
	E. Geological Style	193

Appendix

I.	Polytropic Models	195
	A. Polytropic Envelope	197
	B. Polytropic Sheet	197
	C. Quasi-Polytrope	197
II.	Luminosity	198
	A. Opacity	198
	B. The Photosphere	199
III.	The Escape Flux	200

References 201

Index 203

NOTATION

I. UNITS

SI units and accepted variants are used.

Time	s, day = 86,400 s, a = year = 3.156×10^7 s, ka = 10^3 year, 1 Ma = 10^6 year, 1 Ga = 10^9 year
Length	m, km, AU = 1.496×10^{11} m, parsec = 206,265 AU = 3.086×10^{16} m, Å = 10^{-10} m
Mass	kg
Temperature	K
Energy	J
Density	kg/m^3
Pressure	Pa, bar = 10^5 Pa, kbar = 10^3 bar, Mbar = 10^6 bar
Concentration	parts per 1000 (unless stated otherwise)

II. SYMBOLS

a	radius of body; Al$_2$O$_3$ mass fraction
b	kinematic viscosity temperature variation factor (K)
c	velocity of light, 3.00×10^8 m/s; CaO mass fraction; specific heat

NOTATION

d	depth, of ocean
e	volumetric melt fraction
f	energy flux; FeO mass fraction; flattening factor; probability
g	gravitational acceleration
h	a height; a vertical interval
k	fractionation factor; Boltzmann's constant, 1.38×10^{-23} J/K; K_2O mass fraction; incompressibility ($=1/\chi$)
m	mass; MgO mass fraction; a ratio
m_H	mass of hydrogen atom, 1.67×10^{-27} kg
m'	total MgO + FeO mass fraction
n	number; compressibility factor; Na_2O mass fraction; polytropic index ($=1/(\gamma-1)$); concentration in solution; a ratio
p	pressure; volumetric packing fraction
r	radial co-ordinate, radius
s	SiO_2 mass fraction; number of items
t	time
u	velocity
v	velocity
x	number of items; spatial co-ordinate; a ratio
y	layer thickness; spatial co-ordinate; a ratio
z	spatial co-ordinate, depth
A	area; Rayleigh number
A–G	Earth layer labels (Jeffreys–Bullen model)
BP	boiling point
C	compressibility factor $= \chi_0 \rho_0^n$ ((kg/m^3)^n/Pa); moment of inertia about axis of rotation
E	energy total; Earth
G	gravitational constant, 6.67×10^{-11} N m^2/kg^2
H	specific energy; pressure scale height
I	moment of inertia (usually $\equiv C$); radiation flux (W/m^2)
J	Jupiter
K	thermal conductivity (W/(m K)); adiabatic gas factor
L	luminosity, power output (W); latent heat (J/kg)
L	Moon

NOTATION

$L_{(sun)}$	solar luminosity, 3.9×10^{26} W
M	mass
M	Mars
$M_{(sun)}$	solar mass, 1.99×10^{30} kg
MP	melting point
N	Nusselt number (dimensionless heat transfer ratio)
N	Neptune
P	pressure
R	external radius; gas constant (k/m_H), 8.26×10^3 J/(K kg)
$R_{(sun)}$	solar radius, 6.96×10^8 m
S	Saturn
T	temperature
$T_{(sun)}$	solar photospheric temperature, 5780 K
U	internal energy
U	Uranus
V	volume
V	Venus
W	magnitude of gravitational potential energy
X	hydrogen mass fraction
Y	helium mass fraction
Z	mass fraction of constituents except H, He
α	opacity coefficient; surface melting temperature; solubility coefficient; adiabatic gradient; a ratio
β	ideal gas ratio, $\rho T/P \equiv \mu m_H/k$, $m_H/k = 1.21 \times 10^{-4}$ kg/(K J); melting temperature gradient; opacity coefficient; a ratio
γ	angular momentum per unit mass; coefficient of cubical expansion; a ratio; ratio of specific heats of a gas
δ	a thickness, crust, sublayer, upper mantle
ε	energy per unit mass; energy per unit area; a ratio
ζ	energy per unit mass (of H); escape probability function; a ratio
θ	temperature
κ	opacity (m²/kg); thermal diffusivity (m²/s)
λ	central density
μ	mean molecular mass in amu; molecular ratio

v	kinematic viscosity (m²/s)
ξ	a ratio; transparency fraction; moment of inertia ratio I/Ma^2; Emden radial co-ordinate; molecular density coefficient, 396.4 kg/m³
ρ	density
σ	Stefan constant, 5.67×10^{-8} W/(m² K⁴); finite strength (Pa); Poisson's ratio
τ	time scale
ϕ	Emden variable (dimensionless temperature); a ratio
χ	compressibility (Pa⁻¹)
ω	angular velocity; a ratio
Ω	gravitational potential energy

III. SUBSCRIPTS, ABBREVIATIONS

X, Y, Z	H, He, all others
e	exosphere
c	central value; core; cloud; critical
(inner)	inner core
s	scale, surface, photosphere, SiO_2 fraction
g	gas
ℓ	liquid
f	FeO fraction
m	MgO fraction
(sun)	Sun
(earth)	Earth
(seds)	sediments
i	index
i	interior
j	index
k	index
m	mantle, melting point
amu	atomic mass unit ($=m_H$)
BPD	boiling point at depth

NOTATION

MPD	melting point at depth
SVP	saturated vapour pressure
*	characteristic or selected value
CO_2	carbon dioxide
H_2O	water
′	new value, particular value, root-mean-square value
0	initial, zero-pressure
1–4	Earth layer indices; 4 = core

CHAPTER 1
Prologue

A sample of hydrogen-rich plasma collapses to form our Sun. Some scraps remain to form the planets. A few of the larger pieces are trapped in eternal childhood by their own massiveness. A few smaller pieces throw off their bulky embryonic hydrogen wrappings to start a vigorous geological life of continual adjustment to their inexorable passage through a rich, elaborate phase space. And the story is written in a green black rock you can hold in your hand.

I. INTRODUCTION

Planets are a minor transitory feature of stellar systems.

So it is with the planets of our solar system. Some 5 Ga ago in an outer arm of our youthful galaxy a small, hydrogen-rich gaseous mass, perhaps a little denser than its surroundings, perhaps a little less turbulent, began to collapse under its own gravitational field. The increasingly luminous mass contracted. Behind the inward sweep of the outer envelope were a few gaseous fragments. Soon the central mass, heated by the released gravitational energy, was hot enough for it to become a nuclear furnace. And so it continues for an interval of 10 Ga before rapidly expanding and dispersing again the fragments near it. All that has happened has been the conversion of part of the original hydrogen to helium.

In that 10 Ga interval the planets of our solar system are born, develop and are annihilated.

II. AN ISOLATED GRAVITATIONAL SYSTEM

The planets are "satellites" of the Sun, a typical G2-type star, located in the outer part of the Galaxy in the Orion spur of the Perseus arm, orbiting the galactic centre once every 250 Ma. Presumably the distribution of matter in the Sun's vicinity will have changed greatly during galactic time but there is no evidence to suggest that the solar system is other than isolated. A small amount of matter may be scooped up from the interstellar medium: for example, some of this matter could be the source of cometary material. There is, however, the suggestion (Oort, 1950) that cometary material is obtained from a low density "cloud" of matter, of radius about 50,000 AU and mass of order 10^{23} kg, which travels as a part of the solar system. Also there have been suggestions about injection of exotic material. Nevertheless, the little information we have indicates that the total extra mass scooped up or injected during solar system time is small.

We therefore make the fundamental assumption that the solar system operates entirely from the resources of matter and energy deposited in it at its birth. This resource is a mass of 2×10^{30} kg mainly of hydrogen and helium currently located in a volume of radius 40 AU.

As seen today the matter of the solar system is concentrated in small, compact bodies in an interplanetary medium of extremely low density. There are many such bodies but the bulk of the matter is contained in the few bigger ones. A mere 28 bodies have masses greater than 10^{20} kg. Of these the Sun and the Jovian planets—Jupiter, Saturn, Uranus and Neptune—are the dominant features of the system.

The work of Galileo, Copernicus, Brahe, Kepler and Newton showed that, as a mechanical system, the solar system is driven by gravity. The planets move nearly independently of each other as a "satellite" system around the Sun; satellites move similarly as nearly independent systems around their planets.

III. IDENTIFICATION OF GASEOUS AND STONY BODIES

A fundamental feature of the solar system is that all massive bodies, the Sun and the Jovian planets, are gaseous and the bodies of small mass are stony or icy. This distinction can be demonstrated quite simply using quantities measured external to the bodies. Care is needed, however, to distinguish between the effects of chemical composition and compressibility.

Consider the mass–size relationship for the planets. If all the planets were made of the same material and the effects of compressibility were negligible, then, for example, the mean densities of the planets would all be the same. This is plainly not the case: the Earth has mean density of 5500 kg/m^3, while for Jupiter it is 1400 kg/m^3.

In order to distinguish between the roles of different materials and compressibility we need a parameter which has a dominant effect on compressibility. A suitable parameter is a pressure scale. An appropriate pressure scale is $P_s = \bar{\rho} g a$, where $\bar{\rho}$ is the mean density, g the surface

III. IDENTIFICATION OF GASEOUS AND STONY BODIES

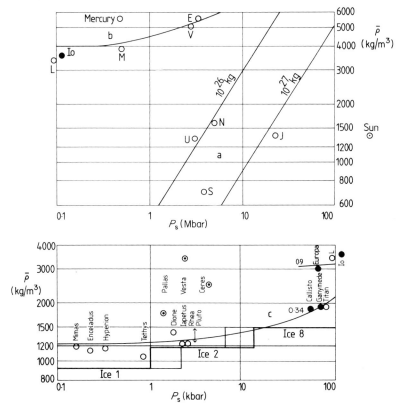

Fig. 1.1. Mean density $\bar{\rho}$ (kg/m^3) as a function of the scale pressure, P_s (kbar, Mbar) for the Sun, planets and their major moons. The lines shown are for polytropic gaseous bodies of mass 10^{26}, 10^{27} kg; Earth-like bodies of various masses; ice–rock bodies of rock mass ratio 0.34, 0.90; and part of the phase diagram of ice.

acceleration and a the radius of the planet. For example, for the Earth $P_s = 3.4$ Mbar, to be compared with, say, the central pressure of 3.6 Mbar. The values for the planets are shown in Fig. 1.1.

The planets and their moons plot in distinct parts of the diagram: (a) the Jovian planets; (b) the terrestrial planets; (c) the icy moons.

1. If the planets were all made of the same material we would expect a monotonic relation $\bar{\rho}(P_s)$. This is not the case, so there is no doubt that the planets are made of different materials.
2. If the planets were incompressible, $\bar{\rho}$ would be independent of mass. This is also plainly not the case. In particular, the Jovian planets are very compressible.

Thus, as we shall see in more detail, there is a clear distinction between the large gaseous planets and the small stony ones, and the role of compressibility needs to be taken into account. A number of lines obtained from simple hypothetical models are shown on the diagram.

Region (a): here are lines of given mass and given radius for a wholly gaseous body. These bodies are discussed in Chapter 3, and details are given in the Appendix.

Region (b): this is for bodies with the same properties as the Earth except for different masses and $m_{(core)} = 0.3$. (See Chapter 7. The mantle layers B, C, D and core materials have the same p_0, n and C values as Earth.) The curve is relatively flat, especially for $P_s \lesssim 0.5$ Mbar, since, compared to a gas, rock substance is weakly compressible. The correlation is quite good, and departures are discussed in detail in Chapter 7: the Moon density is below that of the relation, since its core material zero pressure density is less than the Earth's; similarly, Mercury is above the relation, since it has both a larger core and core material of higher zero pressure density.

Region (c): lines for ice bodies, with a stony central part. Curves for stony mass fraction of 0.34 and 0.9 are shown. Also indicated is a portion of the phase diagram of ice. The bodies in this region are discussed in more detail in Chapter 7.

IV. ON THE ROTUNDITY OF LARGE BODIES

The most obvious structural feature of planetary bodies is their roundness. This arises because of the dominant roles of the two effects of gravity and ease of deformation of matter, whether gas, liquid or solid.

For fluid bodies, departures from sphericity are usually small, arising mainly from the effects of rotation and internal motions; departures from sphericity are restored through the action of compressional waves damped by the viscosity of the fluid—a rapid process. The Sun is very closely spherical; Jupiter is flattened by 6% owing to its rapid rotation.

For solid bodies, the illusion of rigidity arises from our experience of small irregular objects of apparently permanent form. All matter is deformable but the self-gravitational forces producing deformation in a small body are weak.

For solid bodies, the deformability or inelasticity can crudely be represented by the contrasting effects of finite strength and viscosity.

1. Consider a slightly deformed self-gravitating sphere. Then as first noted by G. H. Darwin in 1878 (see e.g. Lamb's "Hydrodynamics",

V. SYSTEM BUDGET

p. 641) there will be flow restoring the body to spherical shape with a time scale

$$\tau = \frac{2(n+1)^2 + 1}{n} \frac{v}{ga}$$

where n is the order of the deformation (of surface wavelength $2\pi a/n$). For $n = 2$, $\tau = 19v/2ga$. This time is geologically short. For example, for typical material of density $3000\,\text{kg/m}^3$ and kinematic viscosity $v = 10^{17}\,\text{m}^2/\text{s}$ for $a = 100$, 200, 500 and 1000 km we have $\tau = 3.6 \times 10^6$, 9.0×10^5 and 3.6×10^4 year. Thus even very small bodies would quickly become spherical.

2. Rock substance, however, has a limiting stress $\sigma \approx 0.1\,\text{kbar}$, as suggested for example by the size and shape of the volcanic pile of Hawaii (Elder, 1976), so that bumps on the surface of a planetary body could be as high as $h = (2\sigma s/\rho g)^{1/2}$, where $2s$ is the base diameter of the bump. This assumes that in the limiting state all the material at the base of the pile is at the limiting stress. Thus $h = (3\sigma s/2\pi \rho^2 Ga)^{1/2} = 89(s/a)^{1/2}\,\text{km}$ for $\rho = 3000\,\text{kg/m}^3$. For example, on a body of radius 1000 km a 200-km wide bump could have a permanent height of 28 km. Clearly, objects of radius of order 10^2 km or less will not necessarily be at all spherical.

In summary: the effects of gravity are completely dominant for bodies of radii larger than a few hundred kilometres, which even if they were severely damaged would return closely to spherical form in times of the order 10^3–10^6 year; nevertheless, bumps and pits of heights typically of order 10 km can be permanent features.

Of the known distinctly non-spherical objects, the largest is Hyperion (Saturn) with gross dimensions $400 \times 250 \times 240\,\text{km}^3$. Also among the striking departures from sphericity are the large craters on the smaller moons, notably the crater Herschel on Mimas (Saturn), which is 130 km in diameter with walls up to 10 km high on a body of only 390 km diameter.

V. SYSTEM BUDGET

The development of a system can be described as a progression through a set of states. The allocation of the amounts of matter, both in quantity and atomic species, and energy and related quantities to the various entities and states of the system, provides a framework for a detailed description. In other words, this point of view regards the essential feature of the development of a system as simply the continual rearrangement of the system resources. Let us make an estimate of the resources of the system in its present state.

A. Atomic Species

The solar atomic abundances are shown in Table 1.1. The elements are listed in order of mass proportion. Notice the following:

1. Only 13 elements account for the bulk of the matter. Table 1.1 includes all elements of mass proportion greater than 10^{-6}.
2. Hydrogen and helium are by far the most common elements. Using the astrophysical notation (X, Y and Z) to represent the mass proportions of (H, He and all others) the tabular values give $X = 0.799$, $Y = 0.184$ and $Z = 0.017$. Measured values of Y are very uncertain and may lie in the range $Y = 0.1$ to 0.4; the value of He abundance in the table has been selected to give $X \approx 0.8$.
3. The Z-component (all constituents except H and He) is very small. Depending on the uncertain value of Y it could lie in the range $Z = 0.01$ to 0.02.

If the system is closed and isolated, and 99.9% of the mass is in the Sun, it is reasonable to assume that the atomic composition of the proto-solar system was identical to that of the present Sun except for the possible effects of nuclear reaction.

B. Mechanical Resources

Consider the planets as a set of particles. Let the orbital radius, mass, rotational moment of inertia and rotational angular velocity of a particle be r_i, m_i, I_i and w_i, with $i = 1, n$, where $n = 9$. Then the contribution to the various quantities of interest for the whole solar system are:

1. Gravitational energy $= -GMm_i/r_i = -mv_i^2$
2. Moment of inertia of solar system $= m_i r_i^2$
3. Angular momentum $= m_i v_i r_i$
4. Kinetic energy $= \frac{1}{2} m_i v_i^2$
5. Rotation energy $= \frac{1}{2} I_i w_i^2$

where v_i is the (Keplerian) orbital speed given by $v^2 = (GM/r)^{1/2}$. These quantities are summarized in Table 1.2.

This shows that the Sun holds the bulk of the mass, rotational and

V. SYSTEM BUDGET

Table 1.1. Solar atomic abundances

Element	amu	Mass (amu)		Oxide (‰)	Symbol
H	1	1,000,000			
He	4	230,000			
O	16	9400	a		
C	12	4300			
Fe	56	1800	a	316	f
Ne	20	1500			
N	14	1200			
Si	28	1000	a	345	s
Mg	24	730	a	296	m
S	32	520			
A	40	300			
Ni	59	120			
Ca	40	86	a	21	c
Al	27	68	a	12	a
Na	23	46	a	10	n
Cr	52	23			
F	19	19			
Cl	35	11			
P	31	8			
K	39	4	a	1	k
Mn	55	4			
Co	59	2			
Zn	65	2			
Ti	48	2			
z		55			

Mass proportions, now, of the elements (with their nominal atomic mass in amu) of abundance greater than 1 ppm, relative to total H = 1,000,000. Value for He uncertain (see text)—the table value gives $X = 0.799$, $Y = 0.184$, $Z = 0.017$. z = all other constituents, each less than 1 ppm. a = major rock constituents, together with their molecular proportions represented as oxides, and the oxide symbols used in the text (the mean atomic mass of these constituents is 24.4 amu, as oxides 58.4 amu). Grand total 1,251,200 amu; total, less H and He, 21,200 amu— to convert to mass ppm multiply values by 0.8. Data from various sources; taken from a compilation by Novotny (1973, pp. 46–49).

gravitational energy, and that the planets hold the bulk of the moment of inertia, angular momentum and kinetic energy.

C. Energy Budget

(The kinetic energy and rotational energy are minor energy contributors and will be ignored.)

Table 1.2. Solar system budget

	Total	Sun	Planets	Ratio Planets/Total
Mass (kg)	2×10^{30}	2×10^{30}	2.7×10^{27}	0.00134
Moment of inertia (kg m^2)	5.1×10^{51}	6.4×10^{46}	5.1×10^{51}	1.00
Angular momentum (kg m^2 s^{-1})	3.2×10^{43}	1.9×10^{41}	3.1×10^{43}	0.994
Energy (J)				
Kinetic	2.0×10^{35}	0	2.0×10^{35}	1.00
Rotational	3.2×10^{35}	2.8×10^{35}	4.5×10^{34}	0.14
Gravitational	-3.0×10^{41}	-3.0×10^{41}	-4.0×10^{35}	1.3×10^{-6}
Nuclear	1.3×10^{44}	1.3×10^{44}	0	0

The major energy requirements are those of the Sun. Suppose it has had the same luminosity for its present life T. Then the energy required per unit mass $\varepsilon = L_{(sun)}T/M_{(sun)} \approx 3 \times 10^{13}$ J/kg. Where is this energy to come from?

1. Chemical energy at typically 2×10^7 J/kg is clearly not a candidate.
2. Gravitational energy, of order GM/R per unit mass, gives of order 10^{11} J/kg. (A factor arising from the particular internal mass distribution, and the factor of 1/2 arising from the consequences of the virial theorem, can be ignored in this order of magnitude estimate.) This is a very useful contribution—indeed a vital one for the proto-Sun. It is enough to run the Sun for of order 10^7 year—referred to as the Kelvin time scale.
3. Nuclear energy is the only viable candidate.

At the temperatures in the central region of the Sun estimated from a variety of models by many workers we expect that the dominant nuclear reaction is the fusion of H to He in the "PPI chain": $p(p,e^+ + v)d$; $d(p,\gamma)^3$He; ^{3}He(^{3}He, $p+p$)^{4}He. The consequent mass loss ratio is $\Delta m = 7.09 \times 10^{-3}$ kg/kg H, with energy $\varepsilon = \Delta mc^2 = 6.4 \times 10^{14}$ J/kg. This is clearly an ample supply. If, for example, during the life of the Sun as a G2-type star, 10% of the mass is converted, there is 6.4×10^{13} J/kg available—a total usage of 1.3×10^{44} J, about 400 times that available from gravitational energy.

This result can be further appreciated by considering the nuclear time scale for an initially pure hydrogen body of constant luminosity, $\varepsilon M_{(sun)}/L_{(sun)} = 103$ Ga. With an anticipated G2 life of 10 Ga, only a fraction of the hydrogen will be converted: the change in the hydrogen mass fraction ΔX in time Δt

is $\Delta X \approx L_{(sun)} \Delta t / \varepsilon M_{(sun)} \approx 0.05$ for the present Sun with $\Delta t = 5$ Ga—a change from, say, $X = 0.85$ to $X = 0.80$ now. The change in the atomic proportions of the solar system have been small and these are confined to the Sun.

VI. OVERVIEW OF IDEAS ABOUT THE ORIGIN OF THE SOLAR SYSTEM AND THE PLANETS

In spite of the vast amount of information and ideas obtained from detailed studies of meteorites, samples of lunar rock, the planets by earthbound and especially space probes such as those of the various Mariner, Viking and Voyager missions, and the Earth itself by geological, geochemical and geophysical techniques which have taken our knowledge back to early in geological time, we still have far more questions than generally accepted answers to how our solar system formed. Many of these questions have been posed in the modern scientific idiom for more than 300 years.

From the early discussion of, for example, Descartes (1644), Kant (1755) and Laplace (1796) two broad themes have been established.

A. Catastrophe

The theme is preoccupied with the origin of the planets. In its simplest form it assumes a pre-existing proto-Sun from which a relatively small proportion of matter was ejected, perhaps because of a near-collision with another star, to thereafter condense into planets. Although at one time these ideas held the centre of the stage they are now resting. A major objection was that the planetary orbits would have large eccentricities contrary to observation.

B. Nebula Condensation

This theme is preoccupied with identifying the processes whereby both the Sun and the planets together coalesced into separate bodies from a presumed single, more or less distinct patch of interstellar matter. This theme is indeed the basis of modern ideas on the formation of stars in general. As applied to the solar system certain apparent difficulties have concerned various workers:

(a) The mass distribution with 0.1% in the planets.
(b) The angular momentum distribution with 99% in the planets.
(c) The mechanism of the production of solid bodies in a predominantly gaseous system.

In my own view these features were (and are) grossly overemphasized, and for too many years directed attention away from several other much more demanding questions. I consider items (a) and (b) to be just what one would expect. Let us consider the apparent problem of the angular momentum by comparison with other systems. A significant fraction of stellar systems are binary. The angular momentum per unit mass of two bodies mutually rotating about their mutual centre of gravity is about $\gamma = 2\pi(a'm')^{1/2}$ AU^2/year, where a' is the orbit radius in AU and $m' = $ (total mass)$/M_{(sun)}$. For, say, $m' = 1$ and $a' = 50$ (typical values) we have $\gamma = 3.2 \times 10^{16}$ m^2/s. Compare this with the value for the solar system of 1.6×10^{13} m^2/s. Thus compared to a typical binary star system the solar system has very little angular momentum—there was no possibility of the proto-solar system splitting into two major fragments.

Item (c) does, however, provide two important extreme sub-themes: (i) direct condensation with possible progressive loss of volatiles during the proto-planet stage, or (ii) condensation first into very small bodies (dust or planetismals) which subsequently accumulate into larger bodies. I should like to note immediately the bias of this book towards sub-theme (i). As I shall show, it is possible to account for the existence of large gaseous planets and small stony planets by a single simple mechanism. Sub-theme (ii) can produce a plausible story for the stony planets, but is then in trouble with the larger gaseous planets (where did all the gas come from?)—I believe this sub-theme fails because of its preoccupation with the stony planets (in any event the bulk of planetary material is gaseous).

Modern ideas are conditioned by a rapidly growing mass of observations about possible stellar birthplaces in the vicinity of our solar system (see, for example, Chapter 4 in Henbest and Marten, 1983). Several possible planetary systems are suggested by some tempting observations. Several current studies have fascinating possibilities but until there is direct evidence of actual proto-planetary systems we must rely on the evidence of the solar system itself. Let me refer to just one aspect of these studies. Studies of luminous blue stars, spectral types O and B, especially in their common grouping, have led to a number of ideas that could have some bearing on the origin of the solar system. These stars are massive, of short lifetimes, and the size of groups is of the order 100 parsec. The best known is the Orion OB1 association. Current models involve a supernova event as a trigger for the formation of the association from a pre-existing nebula. This supernova trigger is an attractive possibility for the solar system in two ways. (1) It provides an "obvious" (?) starting mechanism. (2) It allows injection of material external to the nebula from the supernova—and this exotic material may be the product of extended nuclear synthesis within the proto-supernova. For example, this mechanism has been proposed as the source of the apparent

VI. OVERVIEW OF ORIGINS

excess ^{26}Mg (from the decay of ^{26}Al) found in the Allende meteorite. I do not allow myself to become too excited about these apparent anomalies when I am (we are) still having difficulty in providing mechanisms for such gross features as the occurrence of small stony planets and large gaseous ones.

Although these matters are of great interest, the focus of this book is at one removed. Here my story starts with a collapsing cloud, one which has already become smaller than, say, 10^3 AU. Stars form! The fundamental assumption is made that the solar system was atomically closed early in its pre-stellar stage.

Much of the information has been obliterated during the development of the system, but it is widely believed that there is a clear message "fossilized" in meteoric material early in solar system time. Radioactive dating gives long term ages of 4.5 Ga or so. (Depending on the assumptions made in analysing the data, this figure ranges from 4 to 6 Ga. The oldest rocks so far identified on Earth are meta-sediments of age about 4 Ga. A nominal solar system age of 5 Ga is chosen in this work.) The atomic composition of the non-volatile components of the carbonaceous chondrites is close to that of the present Sun—direct evidence of the proto-solar system composition.

The meteoric message is as yet undeciphered but some parts are clear.

1. The chemical environment was hydrogen-rich. The almost ubiquitous presence of free metal (Fe and Ni alloys) in the whole range of meteorites, including the chondrites, indicates reducing conditions. This is the evidence for the existence of a "hydrogen era" early in solar system time.
2. The structure and texture of meteorites indicate a vigorous physical environment. Observation of the petrographic texture of typical meteorites, both in hand specimen and thin section, suggests strongly (at least to me) that the process of formation was in general far more vigorous than in the case of all terrestrial igneous rocks except perhaps for some of the most violently erupted lavas. One's mental picture is of a moderately dense hot gas filled with vast "thunder-clouds" of whirling droplets, particles and boulders in which the matter is frequently changing phase.
3. The structure and texture of meteorites indicate a wide range of thermodynamic states, especially in local regions where segregation of phases occurred. Consider, for example, the following extremes.

The great bulk of meteorites are chondrites, characterized by millimetre-sized round grains, within many of which are glass and skeletal crystals of olivine and pyroxene, apparently fractured, indicating rapid cooling of droplets of molten rock substance. It was noted by H. C. Sorby (about 1870) that these are similar to the chondrules of slags. His suggestion that they were blown out of the Sun is an unlikely possibility, but the concept of a cloud of droplets being swept about in a field of steep thermal gradient is as good an idea as any.

Some meteorites have a distinct cumulate structure (eucrites); some are constructed of very distinct phases (pallasites); some are even monomineralic (some irons), suggesting formation in a dense body in which fractionation occurred in a liquid melt.

It is especially striking, however, in spite of the wide range of thermodynamic state, that all this material appears to have formed at low pressures—therefore in small localized regions of high density. A few mineral assemblages indicate pressure up to 10 kbar, most indicate less than 3 kbar, and the lack of shape distortion of the chondrules indicates less than perhaps 0.1 kbar.

(There is much speculation about the energy source responsible for those bodies which were melted. Unfortunately the notion that the source is the energy of radioactive decay is still widely held—I cannot take this idea seriously when the available gravitational energy is many orders of magnitude greater.)

In summary, all this suggests a turbulent patchy medium, composed of a hydrogen-rich dusty gas, of small-scale patches, in which local temperatures were high enough for melting to occur, and that this episode persisted perhaps for a time of order 0.1 Ga. This is the environment envisaged for the birthplace of the planets, especially the terrestrial planets. (Undoubtedly meteorites provide the main message from the hydrogen era, but with these broad statements, in this book, apart from a few remarks here and there, I must bid them goodbye.)

VII. VIEWPOINT OF THIS BOOK

The ionized gas-clouds in our neighbourhood of the Orion spur are chemically more or less homogeneous. Plainly this is not so for the present solar system. Just, as it were, look out of your window at the cloudy gas of our nitrogenous atmosphere, at the liquid water running off the snowclad mountains over the rocky riverbed perhaps built in part from originally liquid rock substance. Look at the variety of meteorites. Look at the great balls of hydrogen, the stony planets, the icy moons, the methane ocean of Titan. For my part, fascinating though the clockwork mechanism of our solar system may be, as are the questions of how it came to arrange itself in terms of size, mass and disposition of parts, the most intriguing question is how, from such a

VII. VIEWPOINT OF THIS BOOK

bland material as galactic plasma, could such a variety of phases and consequent structures be built.

Here, then, is the theme of this book. It is simply the view of a sequence of processes of segregation and phase changes through time and the material structures which thereby arise. In this book the word "structure" takes on a broad meaning, referring not at all just to a rigid fixed mechanical entity but to a continually changing thermodynamic state embedded in a form developing in response to the processes of segregation and phase change.

There are many possible processes and states. The phase diagram of the system is very complex. Segregation and sorting processes include: simple self-gravitational collection; the effects of differential flows in multicomponent, multiphase systems, including, e.g., fractional crystallization in a liquid; and chemical processes, e.g. leading to atomic exchange between different phases. The state of individual components may be as simple gas, liquid, solid phases or mixtures such as dusty gas, foam, immiscible liquids and a variety of solid phases. With temperatures ranging from less than 100 K in bodies beyond the orbit of Saturn to 1.5×10^7 K at the centre of the Sun, and matter densities from less than 10^{-7} kg/m^3 at the moment of emergence of the Jovian planets to 10^4 kg/m^3 at the centre of the Sun, the range of possible states is great.

The essential feature, as presented here, of the development of planetary bodies is the progressive separation of the original proto-planetary matter into distinct thermodynamic phases which accumulate in distinct portions of the body. One broad thermodynamic path through the many possible sequences of paths is as follows.

1. *X-gas.* I refer to a wholly gaseous body, presumed here to be of similar composition to that of the present Sun, namely predominantly hydrogen and helium. The most massive bodies, the Sun and the Jovian planets, cannot develop in the sense of this description and remain as X-gas bodies.

2. *Z-gas.* I refer to a gaseous body which has been depleted of most of its hydrogen and helium and other possible volatiles. (When this occurs on a massive scale I shall call the depletion process "H-flushing".) The body may be wholly gaseous or wholly dust but will, in general, largely through the presence of residual volatiles, be a dusty gas. This is the essential stage, as presented here, in the gestation and birth of the terrestrial planets and other rocky bodies of the solar system. Bodies of sufficiently small mass cannot progress beyond this stage, and collapse by accretion without any further major phase change.

3. *Liquid core with Z-gas envelope.* Internal heating during gravitational

collapse can produce a molten central region. This is the proto-geological stage of the terrestrial planets. Whereas in states (1) and (2) the dominant thermodynamic control is through a balance between release of gravitational energy and internal energy and radiation away from the body, a liquid core as it grows progressively becomes the dominant matter and energy sink.

4. *Liquid core with residual envelope (atmosphere/hydrosphere).* This state is presented here as the final pre-geological state of the terrestrial planets.
5. *Liquid core with crust.* This is a molten body which has cooled sufficiently to acquire a thin crustal "skin". The onset of this state signals the beginning of geological time for that planet, since there is then the possibility for the first time for a permanent record of the thermodynamic state to be retained. In states (4) and (5) the residual envelope may develop through a sequence of gaseous, liquid or solid phases as the atmosphere/hydrosphere develops.
6. *Liquid core with crust and mantle.* Further cooling leads to a growing proportion of solid matter to form the mantle as a distinct extract from the pool of matter in the liquid core. The mantle becomes the dominant thermodynamic control on the development of the body. For bodies in which the dominant Z-phase is water substance (as in the many small cold moons of the Jovian planets) stages (5) and (6) have additional degrees of freedom. The liquid core will be water substance saturated and the crust and mantle will be of ice.
7. *A completely solid body.* The body will be inhomogeneous, with more or less distinct crust, mantle and the solidified residual of the final depleted liquid core.

This book has a very strong emphasis on the thermodynamic nature and origin of the core of the Earth and the other terrestrial planets. (The hydromagnetic aspects of the core are not, however, discussed.) It is no exaggeration to say that the key ideas presented here arose from my own puzzling over such simple questions about the core as: of what is it made; is it getting bigger or smaller; how did it form in the first place? As the reader will see, the core forms the heart of the scenario presented here for the terrestrial planets.

This presents a minor problem of an appropriate term for "core" in the broadest sense. Rather than invent another term, in this work I use the term "core" to be not only synonymous with "liquid core"—as for the present Earth, all that material below the mantle–core interface including the mush of the inner core—but also to refer to a fully or nearly fully more or less distinct liquid region in the innermost portion of a body. Thus, for example,

VIII. OUTLINE OF THE BOOK 15

a gaseous dusty body with a central molten zone will have that central zone referred to as the "core", provided that a significant portion of the matter in that zone is molten (it may be as a continuous liquid, a foam or a cloud-like body of liquid drops, together with gas and dust). The "core" in an extreme case could be an entire planet, provided that it was all partially molten and had no gaseous envelope.

VIII. OUTLINE OF THE BOOK

The material is presented in two parts.

Part I, "The Hydrogen Era", deals with those aspects of the system in which a gaseous hydrogen and helium phase plays a dominant role. This of course overlaps in time with the "Geological Era".

A cartoon of the onset of the planets, in Chapter 2, sets the scene. The development and structure of the large hydrogen bodies, the Sun and the Jovian planets, is sketched in Chapter 3. No further direct comment is made about these bodies.

There follows an important Chapter 4 on the initial development and structure of the proto-terrestrial planets. The novel feature is the role and importance of the separation of hydrogen and helium, of amounts of order 10^2 the present terrestrial planet masses, to leave a rock substance residuum. This is the vigorous process I call the "H-flush".

Part II, "The Geological Era", deals with those aspects of the system in which solid phases play a dominant role.

The onset of geological time, in Chapter 5, is seen in the context of the development of the proto-atmosphere–hydrosphere–crust. A simple radiation model is used to describe the development of the system for Venus, Earth, Mars and Titan—all the bodies with an atmosphere.

Data from upper mantle rocks is used to calibrate and identify a model of the development of the chemical structure of the terrestrial planets in Chapter 6. This is an important and key part of the story. It is, as it were, where astronomy meets geology—and they meet very well. Present-day Earth structure data in Chapter 7 allows detailed calibration of physical models of the terrestrial planets. The structure of the ice–stone moons of the Jovian planets is presented here, too. The development of the mantle–core structure for the terrestrial planets is presented in Chapter 8. The three Chapters 6–8 form an interlocking set of ideas about the development of the chemical, physical and thermodynamic structure of the terrestrial planets.

The book is restricted to treating bodies on a global scale. The evidence for the model used here comes in part from our knowledge of the geological development of the terrestrial planets. There is sufficient data on planetary volcanism to allow a partial test of these ideas.

The terrestrial planets receive more attention. Not only do we know more about them but, more importantly, they are much more evolved bodies than are the Jovian planets.

Let me now begin my story.

PART I

THE HYDROGEN ERA

About 5 Ga ago the solar system became an isolated closed system of fixed resources of gaseous matter and energy. After 10 Ma those resources sorted and segregated themselves under the action of the system's own gravitational field into two groupings of gaseous material: the Sun, massive enough from the release of its own gravitational energy to produce temperatures sufficient to start its nuclear furnace; and a miscellany of scraps, the gaseous proto-planets. All these bodies were predominantly composed of hydrogen and helium. And so it largely remains, the Sun evolving along the stellar main sequence, converting hydrogen to helium. The four Jovian planets continue their progressively slower contraction as weak net emitters of radiation. The multitude of bodies of smaller mass, however, the proto-terrestrial planets, moons, asteroidal bodies and meteoritic bodies, during an interval of 100 Ma rapidly flushed away the volatile hydrogen and helium and minor volatiles, leaving partially gaseous and molten residuum of rock substance and water substance to end the first stage of the chemical fractionation of these ultimately partly solid bodies.

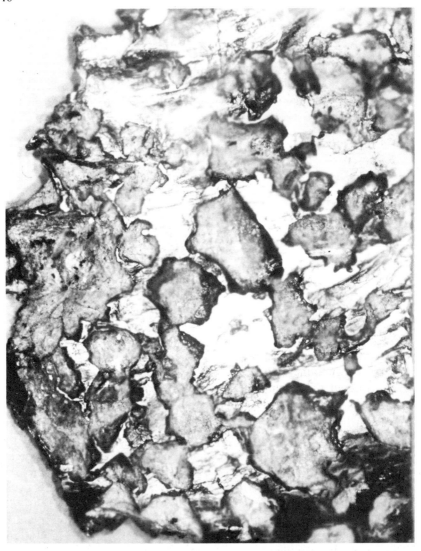

Message from the hydrogen era.

The Imilac (Chile; near 69 W, 24 S) meteorite. A view of a 3 × 4 cm portion of the surface showing a cluster of olivine crystals, dark grains; in a Fe–Ni alloy matrix, light intergrowth.

Classified as a pallasite, a stony-iron. Stony-irons account for about 1.5% of the total number of meteorite falls; of these the pallasites account for a third, namely 0.5% of the total.

(Photograph: author's collection. Specimen in collection of the British Museum (Natural History), London, with permission.)

CHAPTER 2

Gestation of the Planets

I. INTRODUCTION

The solar system starts its life as a single continuous entity. We do not know how this entity came into existence. Intriguing and fascinating though questions about the origin and initiation of the system are, and notwithstanding the multitude of observations and ideas currently emerging from the study of stars and their birthplaces, that is not the subject of this book. Between the covers of this book I am interested in the planets and their global structural development.

Nevertheless, it is not possible to do this without a clear idea of the initial conditions. The bodies of interest progressively change their structures. It is not possible to represent planetary structures as static entities. Rather, one needs to consider a set of structures embedded in a structural space, with the body of interest passing along a particular path in this structural space. There are many possible paths; the particular path is determined by the initial conditions.

The initial system, accepted almost without question in this book, is an isolated patch of gas—the nebula hypothesis. This chapter presents ideas about how particular structural features emerge from that patch of gas. This is done in a qualitative manner by representing the patch as a turbulent collection of mass points.

The main feature of the early structural development is the progressive segregation of the material into a number of isolated portions. A key question then arises as to the nature of the matter in those portions. In this book I make the strong assumption that the matter is an unaltered sample of the original nebula. This viewpoint is very different from that of the accretion/planetismal hypothesis. The chapter closes with an assessment of the direct evidence for these two viewpoints.

II. THE PRELUDE

Current studies of observable interstellar material by means of molecular line and infra-red astronomy suggest that stars are currently forming from "molecular clouds" which are of length scale of order 10 parsec; have differential velocities of order 1–10 km/s; have temperatures of $10–10^2$ K; have densities of order 10^{-18} kg/m^3; are nearly neutral, with relative ionization levels 10^{-5}; and have small magnetic field strength, typically 10^{-6}–10^{-4} Gauss. (See, for example, Mestel, 1977.)

The initial size of the portion of nebula from which the proto-solar system arose could be as great as typical stellar distances, say, of order 10^5 AU. (The instability limit for the solar system, for which perturbations from nearby stars would allow escape of matter, is also of order 10^5 AU.)

The sequence of events during the early stages of the collapse could be thus:

1. Initially the cloud is thin and transparent.
2. The local fragment begins to collapse, perhaps owing to a local fall in turbulence level.
3. Initially opacity is low and radiation loss easy, so the cloud stays cool.
4. Ultimately density and opacity rise, the interior heats up, and the rate of collapse is slowed and controlled by radiation loss from the outer parts of the fragment.

Thus to collapse to the size of the proto-solar system of, say, 10^2 AU immediately prior to the first emergence of the outer planets requires a volumetric ratio and density ratio change of order 10^9. With differential velocities of the order suggested by the measurements, a few km/s, we can envisage a collapsing body with a high level of turbulence. This body will, as a consequence, have been sufficiently homogenized for the particular circumstances of its origin, other than its gross mass, angular momentum and chemical composition, to be obliterated.

III. A TURBULENT SHEET

Envisage, therefore, a nebulous region in a vigorous state of turbulence. A feature of a turbulent gas is the wide range of scales of the matter and velocity distribution. It is very patchy, and within each patch there are smaller scale patches. Observation of the patchiness of the Orion nebula, for example, suggests that in a region of width 10^6 AU there are 10^2 major patches in which stars are forming or have recently formed.

Gravitational effects are stabilizing, thermal effects destabilizing. These patches will be very unstable. Consider, for example, conditions near the margins of more or less spherical patches.

1. Mass, $0.1 M_{(sun)}$; radius, 100 AU; temperature, 100 K. We find $g \approx 6 \times 10^{-8}$ m/s^2; the escape velocity ≈ 1.1 km/s; the molecular velocity (see below) $v = 0.16(T/\mu)^{1/2}$ km/s ≈ 1.6 km/s for monatomic hydrogen.
2. Mass, $0.01 M_{(sun)}$; radius, 1000 AU; temperature, 10 K. The escape velocity ≈ 0.11 km/s and the molecular speed ≈ 0.16 km/s for molecular mass 10.

Such patches will rapidly disperse.

While the number of patches in a region is high so that collectively the region behaves as a continuum, a further process will tend to smooth out the patchiness, namely the net dispersive effect of the region-wide velocity gradients tending to spread a patch by stretching it longitudinally and the diffusive effect of matter escaping from patches acting both longitudinally and transversely. This is the process of dispersion in a turbulent shear flow and is well understood from laboratory studies. The net dispersive effect is related directly to the shear and inversely to the diffusivity. Thus at low temperatures the net dispersion is rapid, because of the dominance of the shear; at high temperatures patches in a shear flow retain identity longer.

The effect of a regional shear flow will be to produce sheet-like bodies, thin transverse to the flow direction. Many nebulae do have such an appearance.

Thus a continual mutual interchange of matter will go on, patches forming and dispersing much as raindrops and vapour do in clouds on Earth.

At some moment, after many rearrangements of the cloud, the mass distribution will by chance arrive locally at a distribution with a slight but distinct permanent central tendency. The patch will very slowly begin to contract. Gravity is winning. This patch of matter has trapped itself. It now progressively becomes distinct from its surroundings and forms an isolated self-contained body. (This description is only a whiff of an answer to such giant questions as to why did the initial collapse start at all and why is the solar system the size it is.)

Let us make a computer simulation of this hypothetical sheet of turbulent gas by considering it built up from patches of material, each patch represented by a group of mass points. Let the points have masses selected at random in some given range. We envisage that each patch remains in existence for a time very short compared to that of the development of the cloud. Let the patches be disposed at random in some given region and continually rearrange themselves subject to three constraints: conservation of mass; the mean angular displacement from one moment to the next is nearly zero, corresponding to nearly zero angular momentum; and the mean displacement is weakly towards the collective centre of gravity. The central tendency will lead to a sequence of stages of structural development, each characterized by the original sheet; a diffuse disc; a central disc; a central concentrated body; and the proto-planetary remnants.

The model is nothing more than a collection of dots slightly organized by a set of apparently innocuous constraints. The model shows a set of structures, a sequence of structures and possible arrangements of the parts of the system. It has no dynamical feature and in particular no quantitative relation to time. The duration of the various stages is discussed in the next chapter.

IV. ONSET OF THE PROTO-SOLAR SYSTEM

Concentrate attention on a region in which a single star might form. Figure 2.1a presents the matter distribution as a set of dots, each representing a unit of mass seen through a 60,000 AU wide window normal to the nebula sheet.

The region is very patchy and contains one dominant patch which has already contracted to radius 3000 AU. This irregular patch continues to slowly contract, continually rearranging its shape but otherwise retaining an irregular form with a weak central concentration. Perhaps an interval of 1–10^2 Ma passes. When the patch is sufficiently contracted for its (very small) angular momentum to require its outer margin to take up that angular momentum, a pronounced structural change commences. A few marginal patches progressively become isolated from the main mass to provide the source of the proto-planets. This stage is shown through the 60 AU window of Fig. 2.1b. The material for the outer planets is already apparent.

Let us now look at these processes a little more closely.

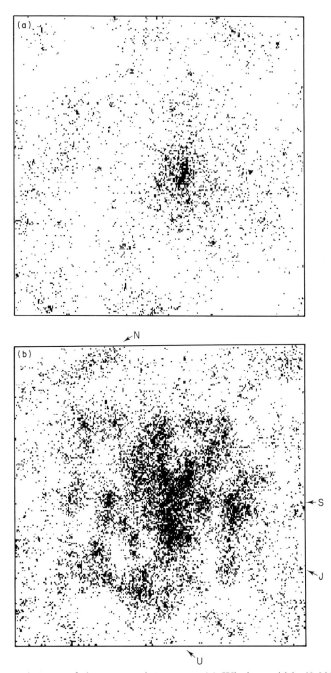

Fig. 2.1. Diagram of the proto-solar system. (a) Window width 60,000 AU. An early view on an intra-stellar scale. Each of the 4000 dots represents a mass of 5×10^{26} kg. The proto-solar system is the dense patch near the centre of the window. (b) Window width 60 AU. A close up of the central patch some time later. Each of the 15,000 dots represents a mass of 10^{26} kg. The material for proto-Neptune, Uranus, Jupiter and Saturn is indicated.

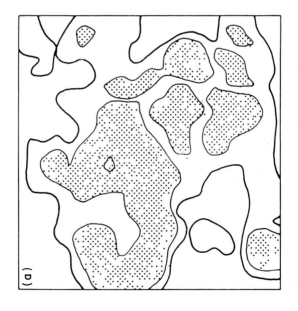
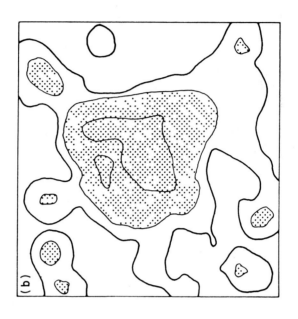

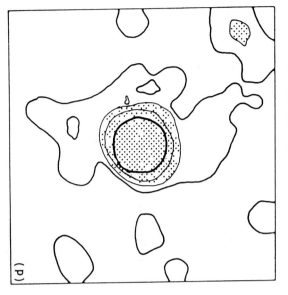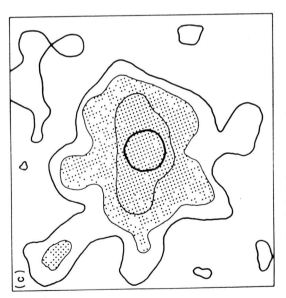

Fig. 2.2. Solar system disc sequence. Contours of areal density in windows of width 2000 AU.

V. EMERGENCE OF THE DISC

The emergence of the disc is illustrated in Fig. 2.2.

The diagrams show a view of the density distribution normal to the sheet, through a window of size 2000×2000 AU, as contours of areal density drawn at $(20, 50, 100, 200, 500, \ldots)$ kg/m^2. During the sequence shown, the emerging and contracting disc changes its radius from 1000 to 200 AU. The development is characterized by the ratios of the disc mass and background mass (that outside the envelope) to the total mass: during the development the disc mass ratio rises from near zero to near unity; the background mass ratio falls to near zero.

(a) Disc mass ratio = 0.08, background mass ratio = 0.92, maximum density ≈ 100 kg/m^2. The central tendency is very weak and barely apparent.

(b) Disc mass ratio = 0.3, maximum density ≈ 200 kg/m^2. A distinct central tendency has developed but the turbulence maintains a strong patchiness of scale 200 AU.

(c) Disc mass ratio = 0.5, maximum density ≈ 350 kg/m^2. The identity of the disc is well developed. The background is already half emptied.

(d) Disc mass ratio = 0.6, maximum density ≈ 1000 kg/m^2. The disc is now distinct with a wispy background. (I shall refer to the outer part of the disc where its density begins to rise above that of the background as the disc front.)

VI. THE ROLE OF ANGULAR MOMENTUM

Figure 2.2 illustrates a scenario in which the proto-solar system develops through a sequence of structural forms: the initial turbulent patchy background; a contracting patchy disc sweeping up the background; a disc which buds off planetary fragments; and the disc itself finally growing a compact central mass which progressively takes up and exhausts the disc material.

Within this scenario the question of why the solar system is the size it is allows us to give a perspective to the role of angular momentum. Consider the following:

(a) The proto-solar system patch is very turbulent. The rate of transfer of vorticity is high and for the mean patch-wide velocity distribution the patch behaves as a "stiff" disc. If at the moment of its isolation the transverse galactic velocity field gradient in the local region is (dV/ds) the patch of initial radius R will have a patch-wide velocity

differential of $\sim R(dV/ds)$. It thereby extracts from the local region an angular momentum contribution $\gamma \sim GM^2(dV/ds)/R$, which as already noted for our solar system is relatively small.
(b) The patch slowly contracts. It remains "stiff".
(c) A distinct central mass concentration grows. The vorticity transfer is still high. The fringe obtains angular momentum from the central region and the mean motion of the fringe approaches Keplerian velocities.
(d) The patch contracts further; the outer remnant, holding the angular momentum it has acquired, remains behind. A small residual angular momentum remains in the central body.

This scenario has features superficially in common with the many models which extract the proto-planets from the proto-Sun. In the scenario here, the proto-planetary material is just left behind, trapped by its angular momentum. During this stage, the contracting disc looks like a body producing "buds" of material—buds which ultimately separate as distinct bodies. The proto-Sun is still a diffuse disc-like body without a concentrated central mass, while the major proto-planets are emerging.

VII. EMERGENCE OF THE PROTO-PLANETS

Out of the patchy background the centrally contracting disc continues to collapse and sweep up the background. Once it is sufficiently small the proto-planetary material emerges. This sequence is illustrated in Fig. 2.3. The diagrams present a window of size 60×60 AU with areal density contours at $(1, 2, 5, \ldots) \times 10^4$ kg/m². (Note that these are merely illustrations—the planets emerge where they do in the diagrams because I have selected a sequence, from many runs of the model, which produces a proto-planet distribution which approximates that of the actual planets.)

(a) Background mass ratio = 0.9, maximum density $\approx 5 \times 10^4$ kg/m². There is a moderate central tendency and no central body, but the field is dominated by the turbulence of scale of order 5 AU.
(b) Background mass ratio = 0.7, maximum density $\approx 10^5$ kg/m². The central tendency has intensified. The material for the Jovian planets is emerging, but is still mixing strongly with that of the central region.
(c) Background mass ratio = 0.5, maximum density $\approx 2 \times 10^5$ kg/m². The material for Neptune, Uranus and Saturn is now quite distinct. The system has a weak spiral structure.

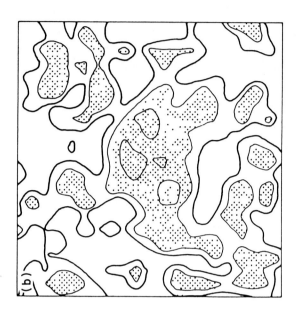
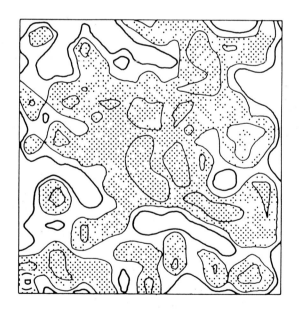

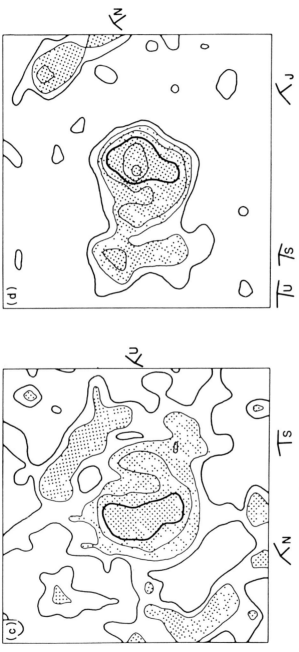

Fig. 2.3. Budding of proto-planetary material during the terminal collapse of the solar system disc. Contours of areal density in window of width 60 AU.

(d) Background mass ratio = 0.4, maximum density $\approx 5 \times 10^5 \, \text{kg/m}^2$. The material for the proto-Jovian planets is largely isolated, although the material for proto-Jupiter is just budding off from the central mass. The background has been thinned to areal density less than $10^4 \, \text{kg/m}^2$. A small concentrated central body is beginning to form within the central patch which is now within 5 AU.

VIII. SUB-PLANETARY MATERIAL

Some material remains which is not now locked up in the Sun, planets and their satellites. This material, derived from residual background material or matter which has had a transitory role in the early life of the major bodies, today constitutes the asteroids, meteoroids, dust and perhaps the comets.

Estimates of the amount of this material now and in the past have a powerful influence on our ideas about the formation of the planets. A commonly held view is that the planets accumulated solely from this material. I have already remarked on this in the introduction. This is the hypothesis of accretion of the planets from a host of cold, solid planetismals. I believe not only that this hypothesis is erroneous but also that its powerful consequences have bedevilled ideas about the early geological history of the Earth and the other planets. The accretion hypothesis is, however, so deeply embedded in the paradigm of this subject that arguments against it are brushed aside. Here I wish to ask the simple question of how much particulate material existed at the beginning of planetary time? We find that there is not enough material.

There *are* numerous small bodies—we collect meteorites and cosmic dust on Earth, some are observed as asteroids or small satellites, or as the ring material of the Jovian planets, and cratering is a solar system-wide phenomenon.

Cratering is seen principally as a chrono-stratigraphic tool. We can also regard the larger bodies as collectors of samples of the particulate material. A major contribution to our knowledge arising from the Apollo missions to the Moon is the dating of rocks in regions of known cratering density. These data pairs tell a simple and very important story of continuity of cratering but of diminishing density with time—a story of a diminishing supply of particles as they are progressively swept up (by the large planetary "vacuum" cleaners).

If the entire proto-solar system was not a cloud of particles, we need to enquire further about the origin of the particles that do exist. The viewpoint of this book, developed in Part I, suggests that within 100 Ma of solar system time all the major bodies have formed. The bulk of the matter remains

VIII. SUB-PLANETARY MATERIAL

largely as hydrogen and helium in the Sun and Jovian planets. All the smaller bodies are fully degassed and the very smallest are not only degassed but remain as primitive unevolved solids. Of these latter we recognize two distinct groupings:

1. Bodies moving in independent orbits about the Sun. This material, with a wide range of orbits, can be crudely divided into "population 1", bodies with very eccentric and oblique orbits extending throughout the solar system, and "population 2", bodies with orbits near the plane of the ecliptic.
2. Bodies trapped in orbits under the influence of major bodies. Some of this material resides in the ring systems of the Jovian planets, and escape from these planets is not possible. A major fraction has been trapped in the asteroidal belt. These bodies have a distinct orbital pattern with several gaps which, as noted by Kirkwood (in 1857), arise from resonances with the orbital motion of Jupiter. I prefer to concentrate attention on the zone between the gaps, for these are regions of trapping. Collisions within the trapping zones produce asteroidal fragments, "population 3" bodies. Collisions continue in the asteroidal belt now as indicated by the few known orbits of meteorites.

A. Particle Mass Estimates

A number of estimates and constraints can be placed on the total particle mass from present-day measurements.

Asteroidal mass. A few thousand asteroids have been detected. Of these the masses of only three are known (units 10^{20} kg): Ceres, 10; Pallas, 2.5; Vesta, 2. As most of the mass is in the larger particles and is dependent on the mass distribution, this suggests a total asteroidal mass of at least 3×10^{21} kg. Thus, if the population 1 mass is taken as including all the asteroids, its mass would have been initially $\gg 3 \times 10^{21}$ kg and would now be $\approx 3 \times 10^{21}$ kg.

Small planetary bodies. The mass distribution for the smaller planetary bodies and moons closely fits $dN/dm = \alpha m^{-\beta}$, where dN is the number in the mass range dm and α, β are constants with $\alpha \approx 2 \times 10^{-3}$, $\beta \approx 0.6$ (for mass in kg). Thus the total mass, $M = \int m\, dN$, of all particles less than mass m, is $M = \alpha m^{(2-\beta)}/(2-\beta)$. The total mass of particles less than 10^{20} kg, typical of the larger asteroids, would be 1.4×10^{25} kg. This is a dubious extrapolation, but taken at face value suggests an upper bound of order 10^{25} kg.

Amount of cratering. Let us estimate the amount of material sufficient to obliterate once by cratering the original surface of the planetary collectors. The number of craters n of average radius r to cover the surface of a body of radius a is $n \approx 4(a/r)^2$. Measurements of particles and craters on Earth demonstrate that (particle radius/r) $\approx$ constant $\equiv \beta \approx 0.1$, so that the average particle mass is $4\pi\rho(\beta r)^3/3$. Let the total collected particle mass be equivalent to a surface deposit of thickness h. Hence we have $h = 4\beta^3 r/3$. For example, with $r = 10$ km, $h = 10.3$ km.

For a total collected mass M over an area A the equivalent thickness $h = M/\rho A$. The collection area is that of the total surface area of all the planets, not just the terrestrial planets and moons, at present 1.28×10^{11} km^2. We note in passing that 99% of this collection area is with the Jovian planets—they have swept up the bulk of the particles. For example, a mass of 10^{24} kg would then give an equivalent thickness of 2.6 km.

Plainly this is enough material to have obliterated the original surfaces 10–10^2 times. Studies of the remnants of the early lunar crust indicate that it has been obliterated, but not necessarily 10^2 times. This suggests a total initial particle mass of 10^{23}–10^{25} kg.

Obliteration will also occur owing to the accumulation of material ejected from craters. The amount of this ejecta is greater than that of the incident particles by a factor of order 10^2. This value is obtained from measurements of artificial explosion pits (Baldwin, 1963) which give the following empirical relations:

1. The radius r of the top of the crater is related to the energy E of the explosion by $E/r^3 \approx \zeta$, a constant $\approx 6 \times 10^9$ J/m^3.
2. The depth H of the crater is such that $H^3/r^2 \approx \eta$, a constant ≈ 16 m.
3. The shape of the crater is approximately paraboloidal, so that the volume ejected $V \approx \frac{1}{2}\pi r^2 H$.

The shape relations also fit those of lunar craters—indeed that is the key evidence that the lunar craters are formed by impact.

Thus for a particle of mass m and impact speed q the kinetic energy $E = \frac{1}{2}mq^2$. Measured impact speeds of terrestrial meteorites range from 10 to 70 km/s (the range for head-on and approaching bodies meeting the Earth at $30(\sqrt{(2)} \pm 1)$ km/s is 12 to 72 km/s), with most being less than 35 km/s. For convenience write $q = eq_0$, with $q_0 = 30$ km/s, the Earth's orbital speed, so that e is in the range 0.3 to 2.4; and take the particle density nominally as 3000 kg/m^3; then the above relations give for the ratio

$$\text{(ejecta volume)/(particle volume)} = Ke^2 r^{-1/3}$$

with $K = \pi\rho q_0^2 \eta^{1/3}/4\zeta \approx 900$. For example, a typical crater with $r = 1000$ m has a volume ratio of 90.

VIII. SUB-PLANETARY MATERIAL

If the original surface was obliterated once solely by craters, there would be such a large volume of ejecta that the ejecta itself would be reworked many times. The estimate of total particle mass required to obliterate the surface is therefore an upper bound—the smaller figure is more likely.

(Some obliteration would also occur where the impacts, especially the larger ones, trigger local volcanism.)

B. Collector Model

Consider therefore the sweeping up of three populations of particles, labelled $i = 1, 2, 3$. The data available to identify these populations is limited, so a grossly simplified model is all that is appropriate. Assume that each population is distributed uniformly in space and that each population has a similar mass distribution. There is no direct evidence for either of these assumptions but there is ubiquitous cratering with similar size distributions on Mercury, the Moon and the larger moons of Jupiter and Saturn; and use of the lunar cratering chronology for Martian geology has not led to any obvious problems.

Let the particle mass scale be m_* so that the total mass of each population $M_i = N_i m_*$ with a grand total $M = \sum M_i$, where the number of particles in each population will be proportional to N_i and the total of all particles to $N \equiv \sum N_i$. Let there be a set of collectors of collection cross-section A_j, $j = 1, 2, \ldots, k$, with the total collection cross-section $A = \sum A_j$. The rate of removal will be proportional to the number of particles remaining and to collection area, so that $dN_{ij}/dt = -\alpha_i N_i A_j$, for some constant α_i, is the rate of loss of particles from population i by collector j.

The total rate of removal for all collectors is then

$$dN_i/dt = d\left(\sum_j N_{ij}\right)\bigg/dt = -\alpha_i N_i \sum_j A_j$$

so that writing $\tau_i = 1/\alpha_i \sum A_j$ we have

$$N_i = N_i(0) \exp(-t/\tau_i)$$

where τ_i is the time scale of population i.

For a particular collector j, the rate of collection of particles $A_j \sum \alpha_i N_i$ gives the cratering rate for that body. For the one body for which we have data, the Moon, total objects for all three populations

$$n = \sum n_i(0) \exp(-t/\tau_i)$$

In converting numbers to masses two important features need to be emphasized. Firstly, although most of the particles are small, most of the mass is in the larger particles. They are the ones which make the visible cratering record. Secondly, from Earth-based cratering studies, we know that the crater size is proportional to the size of the impacting particle. Hence the amount of cratering is directly related to the collected mass and the mass $M_i = \xi n_i$ for some ξ to be determined.

To be explicit, take the particle mass scale for a particle of radius $a_* = $ 1 km, density 3000 kg/m^3, so that $m_* = \frac{4}{3}\pi \rho a_*^3 = 1.257 \times 10^{13}$ kg. A total mass, for example, of 10^{24} kg would then be represented by 8×10^{10} scale particles; and for the estimated total initial equivalent lunar cratering rate of 1.6 crater/km^2, to be discussed, $\xi = 10^{24}/1.58 = 6.25 \times 10^{23}$ kg/(crater/km^2).

C. Identification

The collector model will be identified with three pieces of information: the lunar crater density chronology; a total initial mass upper bound of 10^{25} kg; and a total mass now of at least 3×10^{21} kg.

The data pairs used for the identification here are (time, crater density) with time in Ga and crater density in units of number/km^2 obtained for craters of diameters up to 10 km expressed as equivalent numbers for a common datum of 1 km craters (see Neukum, 1977): 1.0 Ga, 9.5×10^{-3}; 3.3 Ga, 21×10^{-3}; 3.4 Ga, 28×10^{-3}; 3.6 Ga, 33×10^{-3}; 3.85 Ga, 54×10^{-3}; and 4.0 Ga, 70×10^{-3}. The ages lie in a range of at least ± 0.1 Ga; the densities spread over at least $\pm 40\%$. These data are plotted in Fig. 2.4 as density against time (from a nominal start at 5 Ga ago). The fitted curve, together with the contribution from each of the populations, is also shown. The related mass scale is for an initial total (solar system) particle mass of 10^{24} kg.

The three populations for a total initial mass of 10^{24} kg are identified as follows, with the initial mass, time scale and Lunar initial equivalent 1 km diameter crater densities.

1. 913×10^{21} kg, 0.285 Ga, 1.44 km^{-2}. This population is dominant for the initial 1 Ga. A mere 2.2×10^{16} kg remains now.

    The identification is particularly sensitive to the early data values, a factor 2 in crater density chronology leading to a factor of about 10 in the population 1 mass. An extreme interpretation of the existing density chronology requires a total mass of 10^{25} kg.

    The initial rate of removal was 3×10^{21} kg/Ga. It is of interest to compare this with the corresponding rate of accumulation of material from the pre-solar system disc, assuming that took 10^7 year, of about 10^{30} kg/Ga.

VIII. SUB-PLANETARY MATERIAL

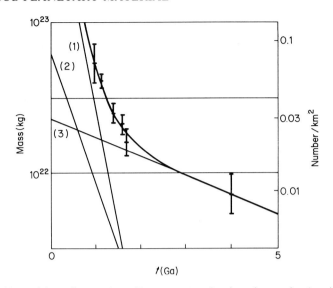

Fig. 2.4. Particle collector data. Lunar crater density, time pairs (vertical bars); fitted curve; components of the fitted curve from populations 1, 2 and 3, given as total remaining mass as a function of time t (Ga). For grand total mass 10^{24} kg.

2. 64×10^{21} kg, 0.505 Ga, 0.10 km^{-2}. This is an important contributor but is never dominant. The amount now is 3.2×10^{18} kg.

3. 23×10^{21} kg, 3.50 Ga, 0.036 km^{-2}. Although the smallest mass contributor this population is dominant after 1 Ga.

The data is inadequate to give much confidence in the identification of population 3. This unfortunately precludes any speculations on the collisional mechanism and rate of production of these bodies.

The total remaining mass now for this model is 5.5×10^{21} kg, almost entirely of population 3 material, to be compared with the estimate obtained solely from the asteroids of about 3×10^{21} kg. Thus 5×10^{23} kg would be a lower bound for the total initial mass. (The present-day Lunar cratering rate is 0.0086 km^{-2}, a fall to 1/180 of the initial rate.)

Of this mass (10^{24} kg) only 1%, the area ratio of the terrestrial planets to that of all the planets, is available for cratering the terrestrial planets, at most 10^{22} kg. This is about 10^{-3} of the total mass of the terrestrial planets (1.2×10^{25} kg). The cratering rates would need to be 10^3 greater than observed for the whole planet accretion mechanism to be possible. There never was enough particulate material.

2. GESTATION OF THE PLANETS

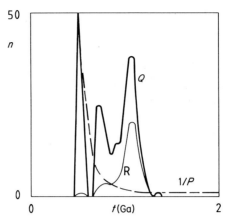

Fig. 2.5. Number of Lunar highland rock samples R as a function of measured time t (Ga). Destruction function is $1/P$ and estimated original equivalent numbers Q.

D. Obliteration of the Early Lunar Surface

Prior to 4 Ga ago the evidence from the cratering record is uncertain. The frequency distribution for samples of rock from the lunar highlands is shown in Fig. 2.5 (data taken from Brown *et al.*, 1977, p. 147). There is a pronounced peak near 1 Ga with no samples before 0.5 Ga. Clearly at least two processes are at work. For times after 1 Ga, cratering persists in the lowlands so that a highland process is terminating. For times before 1 Ga, the lack of samples does not mean that cratering was less or did not exist and that there was a peak of cratering at 1 Ga. It simply means that the early surface has been obliterated by intense cratering.

Consider a process producing material at rate $Q(t)$ subject to partial destruction, such that the probability of survival $P(N) = e^{-N/\lambda}$, where N is the total number of impacts and λ is a constant. If the impact rate is ζn, where ζ is a constant and n the particle rate, the total of impacts after time T is $N = \int_T^\infty \zeta n \, dt$. For the particle populations identified above, only that for population 1 material is of importance during the early interval of interest. Hence $N = \zeta n(0) \exp(-T/\tau)$. The amount left of age T is $R = Q(T)P(N)$. The original amount is then R/P. Figure 2.5 includes the curve of destruction $1/P$ and the resurrected data Q for $\beta \equiv \zeta n_0 \tau / \lambda = 25$. Nonexistent data cannot be resurrected, and the value of β is little better than a guess, but the suggestion that somehow particulate matter was suddenly produced in bulk after a wait of 1 Ga cannot be taken seriously.

Taking the resurrection at face value indicates that for planetary bodies without an atmosphere the first few 100 Ma of the geological record will have been obliterated several times over.

CHAPTER 3

Evolution of the Hydrogen Bodies

I. INTRODUCTION

The original material collects into a number of distinct bodies. All these bodies are chemically similar to the original nebula. They are all bodies composed mainly of hydrogen and helium. The proto-Sun and proto-Jovian planets are sufficiently massive and their exospheres are never hot enough for a significant fraction of their masses to be lost. Throughout their lives they remain as gaseous hydrogen–helium bodies. As a consequence their development is uncomplicated by segregation, fractionation or other phase changes. In essence they are very simple, primitive bodies.

The sequence of events can be described with a model of a contracting disc in which a concentrated central body grows. This disc model is a simple lumped parameter model which, although it ignores the detail shown qualitatively in Chapter 2, allows the identification of time. Furthermore, the disc model can be applied, with suitably chosen parameters for each case, not only to the disc of the proto-solar system but also to each of the sub-system discs of the planets—only the Sun and Jovian planets are considered in this chapter.

II. EVOLUTION OF THE SOLAR DISC

The Sun has two strong influences on its "satellites" during their development. Firstly, the gravity field of the Sun keeps the system together. The strong assumption is made that the orbital arrangement of the planets has been more or less constant throughout solar system time. These matters of solar system dynamics are not the concern of this book. Secondly, the radiation field of the Sun is the dominant influence in setting the temperatures of the photospheres of the planetary bodies. This temperature is a vital ingredient in determining the state of a planetary body. It is of course also affected by the net energy flux from the interior of the body, which is significant for all the proto-planets and is significant today for the Jovian planets. Nevertheless, knowledge of the solar luminosity and its variation in time is the one essential piece of information about the Sun vital to an understanding of the development of the structure of the planets.

The Sun spends the bulk of its time in one of two distinct states.

Until the disc-like patch of the original nebula has contracted to much less than 100 AU, the proto-Sun does not exist. Ultimately a central mass concentration develops within the disc, and the proto-Sun is born and grows, taking up material from the disc, passes its peak size and only then as a distinct entity enters the first major state, one of gravitational contraction.

From the moment when the proto-Sun becomes a more or less distinct body, as a first approximation, it is possible to represent its evolution with a model of a spherically symmetric gas ball, initially collapsing under its own gravitational field until its interior is hot enough to allow hydrogen fusion and the commencement of the second major state. Then for a period of order 10 Ga, during which some of the hydrogen is used up, the Sun behaves like most ordinary stars—i.e., lies on the "main sequence"—before blowing up into a large, bright sub-giant star.

It is sufficient for the purposes of this book to have a clear idea of the main processes involved and the time scales of the major states.

A. Energy Considerations: Link to Time

The construction of a model describing the development of the system of interest will usually have two distinct aspects. Firstly, one envisages a set of possible structures through which the model may pass. Various paths through this structure space may be identified to produce a possible sequence of structures. The range of possibilities will usually be great. Secondly, one attempts to relate a particular structural sequence to time. In detailed models such as those found in continuum mechanics the relations of conservation of mass, momentum and energy, and various constitutive relations, are applied at every point of the structural field. Where, however, as in most of the models of this book, there are insufficient data to calibrate elaborate models, the conservation relations are applied to the model as a whole. Each stage of the sequence is characterized by its energy content. In order to

II. EVOLUTION OF THE SOLAR DISC

change to the next stage, with usually a different energy content, some energy needs to be transferred elsewhere. The power of this transfer process L, together with the amount of energy change ΔE, determines the time interval Δt between the stages from $\Delta E = L \Delta t$.

The systems of interest carry their energy in two forms: gravitational energy arising in their own gravity field; and thermal energy—nuclear energy is referred to in Section VC, Chapter 1. For these systems, matters to do with the energy budget are straightforward. Consider a self-gravitating body of mass M and typical dimension R, say a radius or thickness. Let $W > 0$ be the work needed to disperse the body to some reference state, say infinity. Then $W = \alpha GM^2/R$. (The ratio α is close to unity: 4/3 for a uniform disc; 0.6 for a uniform sphere; 0.88 for a self-gravitating gas sphere.) The gravitational potential energy $\Omega = -W$. Note that this is a negative quantity. Thus, for example, if a body of given mass contracts, its potential energy becomes more negative; it loses potential energy of amount ΔW. The thermal energy U is readily shown to be $U = \frac{1}{2}W$. Thus a contracting body of given mass increases its thermal energy by amount $\frac{1}{2}\Delta W$. Hence there is an amount of energy $\frac{1}{2}\Delta W$ to be removed elsewhere. If the power emitted from the body, the luminosity, is L then the time required $\Delta t = \frac{1}{2}\Delta W/L$. (These relations in the form $2U + \Omega = 0$ are a version of the Virial theorem.)

B. Disc Model

Consider a contracting disc-like body of gas as sketched in Fig. 3.1. The disc will be assumed to be a patchy turbulent body modelled as an areally homogeneous body. At some stage a central compact body will grow and take up the disc mass. We are interested in the temporal development of this disc system as a model of both the proto-solar system and individual proto-planetary systems.

Choose units: length, R_0; mass, M_0; energy, $W_0 \equiv GM_0^2/R_0$; temperature,

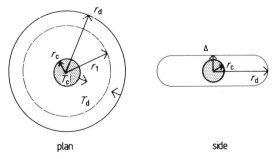

Fig. 3.1. Disc model schema.

3. EVOLUTION OF THE HYDROGEN BODIES

T_0; luminosity, $L_0 \equiv 4\pi R_0^2 \sigma T^4$; time, $\tau_0 \equiv W_0/L_0$. Use subscripts: d for the disc, c for the central body. Thus the dimensionless disc radius $r_d = R_d/R_0$, half thickness $\delta = \Delta/R_0$ and temperature $\theta_d = T_d/T_0$ etc. The total mass $m = m_c + m_d$; the total work function $w = w_c + w_d$; the total luminosity $l = l_c + l_d$ and $\Delta t = \frac{1}{2}\Delta w/l$.

At some time $t = 0$, let the disc radius r_d be r_0. The disc contracts. Three stages will be recognized, characterized by the disc radius. Let the central body begin to form when $r_d = r_1$; and the disc be exhausted when $r_d = r_2$, when also $r_c = r_2$.

1. $r_d > r_1$, disc alone, no central body.
The budget is:
(a) disc: $m_d = 1$, $w_d = \alpha_d/r_d$, $l_d = \frac{1}{2}r_d^2\theta_d^4$;
(b) central body: $m_c = 0$, $r_c = 0$, $w_c = 0$, $l_c = 0$.
The energy equation integrates directly to give

$$t = \tfrac{1}{3}(\alpha_d/\theta_d^4)(r_d^{-3} - r_0^{-3})$$

Thus, if $r_0 \gg r_d$, the time, in units of τ_0, to reach radius r_d is

$$t \approx \alpha_d/3\theta_d^4 r_d^3$$

2. $r_2 > r_d > r_1$, disc with central body.
Assume that the central body begins forming at disc radius r_1; thereafter the disc continues to contract but with constant areal mass density $\rho_1 = M_0/\pi R_1^2$; and that the mass lost from the disc, now taken as an annular cylinder, supplies the growing central body. Then

$$m_d = (r_d^2 - r_c^2)/r_1^2 \qquad m_c = 1 - m_d \qquad r_c = r_* m_c^s$$

where r_*, s are given (see below). For given r_d, say, these relations are readily solved iteratively for m_d, m_c, r_c.
The budget is:
(a) disc: $w_d = \alpha_d(r_d^3 - r_c^3)/r_1^4$, $l_d = \frac{1}{2}(r_d^2 - r_c^2)\theta_d^4$;
(b) central body: $w_c = \alpha_c m_c^2/r_c$, $l_c = kr_c^2\theta_d^4$, $0 \leq k \leq 1$.
This k is a factor determined by the areal fraction of the central body which is not obscured by the surrounding disc. The form

$$k = \left(\frac{\zeta}{1-\zeta}\right)\left[\left(\frac{r_c}{\delta}\right)^2 - 1\right] \quad \text{for } r_c > \delta$$

or $k = 0$ where $\zeta = \delta/r_2$ such that kr_c^2 rises from 0 at $r_c = \delta$ to r_c^2 when $r_c \gg \delta$, provides a suitable approximation.

For a solid body there is a direct relationship between size and mass. Such

II. EVOLUTION OF THE SOLAR DISC

is not the case for a gaseous body—a body of given mass can have any size at all. In a full dynamical model the relationship $r(m)$ will usually be found by integrating along some evolutionary path. Here, in this lumped parameter model, a particular relationship must be imposed. The choice $r \sim m^s$ gives various possibilities: $s = 2$, constant potential energy; $s = 1$, constant temperature distribution, in particular constant central temperature; $s = 1/2$, constant central pressure; and $s = 1/3$, constant mean density and central density. For the illustrations here I have arbitrarily taken $s = 1$.

The temperature of the photosphere of the central body is set by the thermal structure of the body, in particular by the net loss of energy from its interior and the opacity of the material near the photosphere. This needs to be determined at each step of the model—details are given below. We note, however, that for a particular structure and material, here a self-gravitating gas sphere and adiabatic perfect gas, that $T_s = T_{s0} m^p r^q$, where p and q are parameters to be determined; for example, for the proto-Sun $T_{s0} \approx 4000\,\text{K}$, $p \approx 0.24$ and $q \approx 0.04$. In practice, the quantities T_{s0}, p and q can therefore be determined separately once and for all to provide a very simple implementation.

A similar approach could be taken for evaluating the disc photo-surface temperature T_d. This would, however, merely add further unknown parameters to the model. I have therefore taken T_d as a phenomenological parameter.

3. $r_d < r_c$, *central body alone, disc exhausted.*
The budget is:
(a) disc: $m_d = 0$, $r_d = 0$, $w_d = 0$, $l_d = 0$;
(b) central body: $m_c = 1$, $w_c = \alpha_c/r_c$, $l_c = r_c^2 \theta_c^4$.
The energy equation integrates directly as in (1) above to give

$$t - t_2 = \tfrac{1}{6}(\alpha_c/\theta_c^4)(r_c^{-3} - r_2^{-3})$$

Thus the time, in units of τ_0, to reach radius r_c is $\Delta t = \alpha_c/6\theta_c^4 r_c^3$ (note an extra factor of 0.5 compared to the disc). For example, to reach $\theta_c = 1$, $r_c = 1$ requires $\Delta t = \alpha_c/6$.

Role of model parameters. An illustration of model behaviour and the role of some of the parameters is shown diagrammatically in Fig. 3.2. Each of the three stages is indicated: (1) disc alone—the level of this line is set solely by T_d; (2) disc with central body—here there is a wide range of possibilities, the major ones arising from when the central body breaks through the disc and from the ultimate central body size; and (3) central body only—the form of this curve is the same for all bodies, its duration being set by the central body maximum size and photosphere temperature. The illustration shows the role of r_1 and r_*.

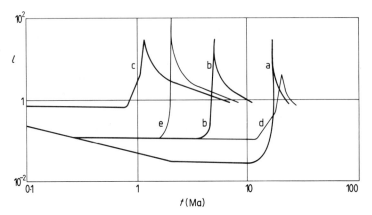

Fig. 3.2. Effect of parameter variation for disc model: total luminosity ratio l as a function of time t (Ma). (1) Curves (a), (b) and (c), $r_* = 2, 5, 10$. (2) Curves (d), (b) and (e), $r_1 = 50, 100, 200$. The reference curve (b), also shown in Fig. 3.3, has $r_1 = 100$, $r_2 = 2$, $r_* = 5$, $\zeta = 1$, $T_d = 200$ K, $T_0 = 4000$ K.

The diagram emphasizes the wide range of behaviour. Furthermore, in the case of the Sun there are no appropriate data with which to calibrate the model during stages (1) and (2): all this information is completely obliterated once the proto-Sun becomes a star. For the Jovian planets, however, there is a little information to be gleaned, since the early stages occupy a significant part of the time of their existence.

An aside. It is of interest to contemplate the consequences of a model involving only stage 3. Models focus attention on a few aspects of a system's nature. Taken to extremes they can give silly results but in so doing force attention on an aspect of the system often previously ignored. Consider the self-gravitational collapse of a body, here the basis of models of all the hydrogen bodies. Since the photosphere temperature of a gaseous body of given mass is very weakly dependent on its size, a good approximation is to take the photosphere temperature as constant and then obtain $t/T = r^{-3}$, where T is a time scale and $r = R/R_0$. For the Sun, $T \approx 4$ Ma, the Kelvin time. This tells us that the gravitational collapse stage takes about 4 Ma—current evidence is compatible with this value. In a little more detail, the time to reach $r = 5$ is 0.03 Ma, and to reach $r = 2$ is 0.5 Ma—thus the collapse slows down and the bulk of the time is spent at radii close to the final one. Again this is satisfactory. Consider further, then, the time to reach $r = 10$, 4 ka; $r = 10^2$, 4 year; $r = 10^3$ (about 5 AU, at Jupiter's orbit), 1.4 day; and $r = 10^4$ (about 50 AU, somewhat larger than the present solar system), 2 minutes (!). There is something amiss—a body falling from infinity towards the Sun would pass from $r = 10^4$ to $r = 10^3$ in about 20 year, not 1 day. How

II. EVOLUTION OF THE SOLAR DISC

would we ever arrange the matter to form the solar system if we only had 2 minutes in which to do it! Clearly, if we are to keep the collapse model it must be modified. There needs to be a further ingredient. Plainly the simple concept of continual collapse is nonsensical. If the body has not always been collapsing there must have been a time in which it was growing. Thus we are led to consider the proto-Sun as starting its life as a mass concentration growing within the nebula, taking its mass from the nebula. This central concentration may remain small—even $r = 100$ is unlikely, and indeed it is likely that at most $r \lesssim 10$. We reach a new viewpoint. The proto-Sun is not the nebula; it is a small body which at a late stage grows within the collapsing nebula until the rate of supply of mass from the nebula cannot maintain the growth, and only then does the proto-Sun begin to contract and enter its self-gravitational collapse phase.

The three-stage model is also inadequate for small times. It presumes that the original patch had a lifetime sufficiently large for the contraction to be established. Thus if the patch lifetime were somewhat less than 1 Ma, no permanent central contraction would be possible. In a fuller model, the process of patch accumulation and dispersion would be represented also. For the model here the early times are not well defined, with uncertainty of perhaps 0.1 Ma.

C. Disc Development

The model temporal behaviour of the disc (for one choice of parameters) is shown in Fig. 3.3.

1. $0 < t < 0.2$ Ma. There is a disc but no central body. The disc collapses from radius greater than 1000 to 100 AU: the disc luminosity falls from greater than 3 to 0.03 (the luminosity being determined by the disc size and the assumed disc photosphere temperature of 200 K).
2. 0.2 Ma $< t < 6.8$ Ma. The disc continues its collapse. A central body begins to form, very slowly at first, reaching $r = 1$ at 6.2 Ma and $r = 5$, its maximum, at 6.8 Ma—when the disc radius is 10 AU. The central body photosphere temperature rises but its effect is muted while the central body is shrouded by the disc. The luminosity rises slowly at first and then rapidly as the central body builds out from the disc. The luminosity peaks at 32.
3. 6.8 Ma $< t < 7.3$ Ma. The disc collapses rapidly and is finally exhausted. The central body begins to contract, rapidly at first. The central body photosphere temperature falls slightly.
4. 7.3 Ma $< t$. The disc is gone. The central body continues to contract, progressively more slowly, its radius falling below $r = 1$ at 11.5 Ma.

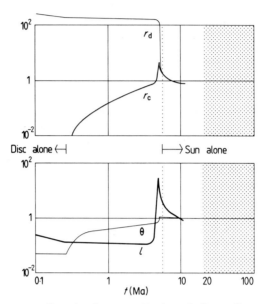

Fig. 3.3. Disc, proto-Sun development. Ratios of disc radius r_d; central body radius r_c; total luminosity l; central photosphere temperature θ as functions of time t (Ma) ($r_1 = 100$, $r_2 = 2$, $r_* = 5$, $\zeta = 0.1$, $T_d = 200$ K, $T_0 = 4000$ K). In the shaded region nuclear processes are dominant and the disc model is no longer valid: the proto-Sun has become a star.

Thus the system has made a transition between two different structures: from a simple disc to a simple spherical body, in an interval of about 5 Ma. After a total time of some 10–20 Ma the contraction of the proto-star comes to an end and the smallest radius of about $0.8 R_{(sun)}$ of the proto-star is reached as the nuclear furnace comes on; within 50 Ma the nuclear furnace is fully lit and the new star is on its slow path up the main sequence.

This episode of solar system history is of interest in this book insofar as it reveals the setting in which the proto-planets begin their existence. The central body, the Sun, is not the focus of our attention.

We reach a turning point in our story. We turn out eyes inward, away from the stars, away from the Sun: we look at the planets themselves.

III. EVOLUTION OF THE JOVIAN PLANETS

After 10 Ma of solar system time the solar disc has collapsed and the proto-Sun is a small, compact body, nearly a star—a few patchy remnants of the disc are organizing themselves into separate systems. Each of these remnants develops into a proto-planet. Here each such remnant is considered simply

III. EVOLUTION OF THE JOVIAN PLANETS

as a miniature version of the solar disc. Each remnant is modelled as a mini-disc. Being smaller and cooler, their time scales are greater: 30 Ma for the solar disc, and 34 Ga for the Jovian discs. Following the prelude, the proto-Sun emerges from its disc after 2 Ma, and the proto-Jovians from their discs after 30 Ma; the collapse stage of the proto-Sun took 10 Ma, while the Jovians are still collapsing.

The four large hydrogen–helium bodies Jupiter, Saturn, Uranus and Neptune have masses of order $10^{-3} M_{(sun)}$. As a consequence, the central temperatures of these bodies are too low to allow nuclear reactions to occur—the limiting stellar mass is about $0.1 M_{(sun)}$. Their evolution therefore is determined solely by their gravitational collapse. Their structures, apart from a change of scale, are similar to what they were early in solar system time.

We look first at the development of the Jovian bodies—the terrestrial planets have their turn in the next chapter—and then briefly at their internal structures as they are today.

A. Emergence of the Proto-Jovian Planets

The disc model will now be applied to the Jovian systems. The model is as before except that the units of length, mass etc. are the present radius, mass etc. of the particular planet. The same three stages are passed through: disc alone, disc with emerging central body, and central body alone.

The case of the Jovians is more fruitful than that of the Sun, which has passed into a different state obliterating any early information. The Jovians are still collapsing and therefore carry some information from the pre-planetary discs.

The total time of development of the Jovian planets arises from the short time spent in the disc stage and the long time of collapse from when they were at their peak. This latter time is about $\tau \equiv 1/6 \alpha_c \tau_0$, where $\tau_0 = GM_0^2/4\pi\sigma R_0^3 T_0^4$ and M_0, R_0 and T_0 are the present mass, radius and temporal mean photosphere temperature. For an age $\tau = 5$ Ga we require $\tau_0 = 34.1$ Ga. This result applies to all the Jovian planets. Hence we obtain an estimate of T_0: Jupiter, 170.4 K; Saturn, 106.8 K; Uranus, 77.7 K; and Neptune, 88.3 K.

A second set of photosphere temperatures can be obtained from the model of a Jovian planet as a self-gravitating body (see Appendix). We find values of T_0 now: Jupiter, 156.8 K; Saturn, 116.5 K; Uranus, 71.9 K; and Neptune, 74.8 K. This is very nice. These values differ by no more than 15% from those estimated from the age of the Jovian planets. Thus, although there are uncertainties in the opacity, in the use of a constant T_d in the disc model, and in the possible integrity of a disc at all, the consistency of these two data sets gives some confidence in the model.

We can now go one step further. Experimenting with the model shows that, for given τ_0 and T_0, the gross development is most strongly affected by r_1 and T_d. Thus we may vary these two quantities and find combinations of values which give an age of 5 Ga. This is shown in Fig. 3.4 as curves of given T_s. Note that for a particular curve there is a critical value of T_d above which no solution exists—the disc cools too quickly to give the required age.

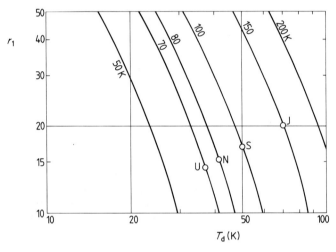

Fig. 3.4. Parameter selection for disc model of the proto-Jovian planets. Disc radius r_1 for onset of growth of the central body, as a function of disc photosphere temperature T_d (K), as curves of given central body photosphere temperature scale T_s (K), for bodies of present size after 5 Ga ($r_* = 5$, $r_2 = 2$ fixed). The selected parameters are indicated.

There is a wide range of possibilities with no obvious criterion for selection. Quite arbitrarily I have selected a set of T_d proportional to (orbital radius)$^{-1/2}$, somewhat above the critical values, but otherwise for small values of r_2, obtaining $T_d = 70$, 50, 37 and 42 K at $r_1 = 20.1$, 16.8, 14.4 and 14.7. We cannot take the numerical values seriously, but they strongly suggest a fall of proto-planetary disc temperatures, as we would expect, in moving to the outer parts of the remnant of the proto-solar disc. The somewhat higher photosphere temperature for Neptune's disc compared to that for Uranus could arise from the fluctuations expected in the turbulent field—but that would be reading too much into the numbers.

The temporal development of the Jovian bodies for these parameters is shown in Fig. 3.5. They are remarkably similar patterns.

III. EVOLUTION OF THE JOVIAN PLANETS

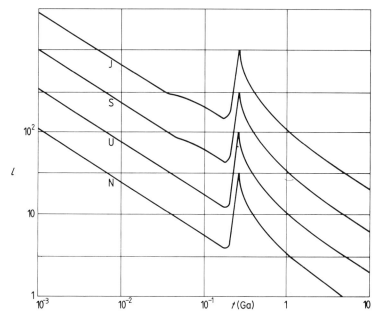

Fig. 3.5. Luminosity history of the Jovian planets. Luminosity ratio l as a function of time t (Ga). The ordinate labels are for the Neptune curve; the other curves are displaced upwards in steps of half an order of magnitude. Jupiter's curve is identical to that in Fig. 3.6. The curves are nearly indistinguishable at the scale of the diagram.

B. Development of Jupiter

A set of model curves for Jupiter is shown in Fig. 3.6. We have the following events:

1. $0 < t < 34$ Ma. There is a disc but no central body. The disc collapses from radii greater than $100R_J$ to $20R_J$; the disc luminosity falls from 240 to 9.6 (level set by assumed $T_d = 70$ K).
2. 34 Ma $< t < 270$ Ma. The disc collapses slowly to $5R_J$. A central body begins to form and grows to its maximum radius of $5R_J$ and luminosity 32.
3. 270 Ma $< t < 770$ Ma. the central body starts its contraction, reaching $2R_J$ as the disc remnant becomes exhausted after 770 Ma.
4. 770 Ma $< t$. The disc is gone. The central body continues to contract, reaching radius $1R_J$ in 5 Ga; the luminosity falls from 4.4 to 1.0.

C. Physical Structure of the Jovian Planets

The information about the Jovian planets as structures is limited. They are large balls of gas. The mass, size and shape are known; but even the moments

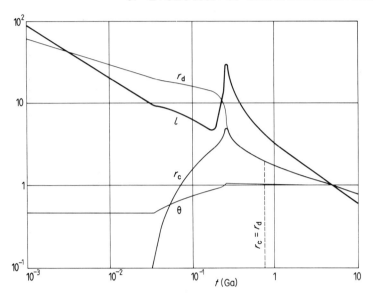

Fig. 3.6. Disc, proto-planet development of Jupiter. Ratios of disc radius r_d; central body radius r_c; total luminosity l; central photosphere temperature θ as functions of time t (Ga).

of inertia are only poorly known. The outermost layers of Jupiter and Saturn are marvellously vigorous—we may guess this is also the case for Uranus and Neptune—current observations giving quite a detailed picture of their chemistry and meteorology. From the interiors there are two messages: there are internally generated magnetic fields, suggesting a vigorous mobile system; and a net outward heat flux several times that provided by the exterior solar radiation, the evidence that the bodies are still contracting. Otherwise our knowledge of the interiors is nil. A number of elaborate models have been produced, involving unnecessarily extreme assumptions and which for the foreseeable future cannot be calibrated. I consider the simple polytropic gas model as providing as good a picture as the existing data warrants. The structure of the interior of these bodies remains hidden from us.

Structure. The internal structure of the Jovian planets represented as polytropic gaseous bodies is illustrated in Fig. 3.7. A particular function is of the same form in all the planets; the only difference is in the length and mass scales. Furthermore, a particular function for a particular planet has the same form through time; the only difference is in the change of length scale, the radius $R \approx R_0(t/5Ga)^{-1/3}$.

Evidence to check the validity of this model, indeed any such model, is limited to some dubious estimates of the moment of inertia, usually obtained as that about the polar axis C.

III. EVOLUTION OF THE JOVIAN PLANETS

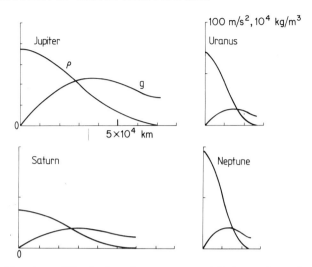

Fig. 3.7. Structure of the Jovian planets now, modelled as polytropic gas spheres; density ρ (kg/m^3) and gravitational acceleration g (m/s^2) as a function of radius. The diagrams are all to the same scales.

For a rotating body, the body form to a first approximation is that of an ellipsoid of flattening $f = f(m, C)$, where $m = a^3\omega^2/GM$ is a measure of the dimensionless ratio of centrifugal to gravitational acceleration. (See, for example, Cook, 1980, Chapter 8, from which the values below are quoted.) Thus from measured values of f and m we can derive C. Unfortunately for the Jovian planets, there is a rather wide range of measured values of f so that C is rather uncertain:

Jupiter: $m = 0.0881$, with quoted measurements $f = 0.060$ to 0.066 and hence $C/Ma^2 = 0.23$ to 0.26.
Saturn: $m = 0.1657$, with $f = 0.088$ to 0.109 and $C = 0.15$ to 0.22.
Uranus: no agreed value.
Neptune: no agreed value.

These estimates can be compared to that for a polytropic gas model for which $C/Ma^2 = 0.21$. Given the wide range of the estimates, it is doubtful if there is anything to be gained other than to note that the values are in fair agreement. With the existing data, the simple model of a chemically homogeneous polytropic gas is adequate.

Between the covers of this book there is no space to linger. While the great gas balls continue on their elephantine way, indeed even before they have commenced their contraction stage, a number of smaller sprightly bodies have already emerged and undergone a dramatic change. The terrestrial planets are forming.

CHAPTER 4

Parturition of the Proto-Terrestrial Planets

I. INTRODUCTION

After the passage of the disc front, the proto-planets have emerged as distinct entities. The larger bodies are sufficiently massive to retain their identities for the remainder of solar system time. Near the proto-Sun, the proto-terrestrial planets are also entirely gaseous bodies composed largely of hydrogen and helium. Their initial individual masses are of order 10^2 times their present value and radii are of order 10^3 times bigger. The temperatures of their surfaces are high: they are collapsing rapidly and the rate of conversion of gravitational energy is high; they are near the proto-Sun, which is still in its very luminous gravitational collapse stage. In a most dramatic sequence of phase changes, escape of the light elements, mainly hydrogen and helium and minor volatiles, occurs easily to leave a dusty gas composed of the rock substance residuum which becomes sufficiently hot as it collapses to produce a molten proto-planet.

Three distinct stages thus occur during the birth of a terrestrial planet. The chemical segregation, early in the life of the planetary mini-disc, produces a dusty gas disc. This stage is described in the first part of this chapter. Some other applications of the ideas are given here, too. Within the depleted disc a central body grows. When the disc is exhausted, the central body reaches its peak size and then begins to contract. It is still composed entirely of a dusty gas. As it contracts its interior becomes hot; ultimately melting occurs in a central region and a growing core of liquid rock substance accumulates until the entire body is molten. These final stages of the parturition are described in the second part of this chapter.

This chapter, and Part I, closes with a chronological summary of the early solar system structural history as described in this part of the book.

II. DEGASSING

If the proto-terrestrial planets commenced their existence as a sample of the nebula, as proposed here, as do the Jovians, they would also contract and ultimately form central gaseous bodies. Clearly something else must have happened in the meantime.

When we contemplate the gross features of the solar system we remark on various outstanding aspects—e.g. Bode's relation between planet number and orbital radius. Surely, however, the most striking feature is that the planets are in two distinct groups: small stones near the Sun; large gas balls at a distance. Surely this is not a fluke. Could it have been the other way around? As we shall see, the answer is a definite no.

Let me put this point another way. There is a large mass gap between the terrestrial and Jovian planets. Why are there no planets of mass intermediate between that of Earth and that of Uranus?

These and other properties arise from the operation of a single process. This process is dominant in the inner part of the early proto-solar system. It may be negligibly weak or, in not very different circumstances, exceedingly strong—the sensitivity to the circumstances being so great that the process is, in effect, on or off. I am referring to the loss of matter from a body simply because the atoms and molecules in its outer layers are hot enough and fast enough to escape from the body.

This process produces the most dramatic rearrangement of matter in the entire history of the solar system (other than its formation and subsequent dispersal, which is strictly a stellar process). The gas balls of the proto-terrestrial planets lose about 98% of their masses to become small, dense bodies of rock substance.

It is important to appreciate that the scenario presented here about degassing is different from that discussed by others. A great deal of work has been done on the escape of species from the atmospheres of the planets. These studies refer to the escape from the planets more or less as they are now; the mass fluxes are minute and the effect on the planets' mass negligible. What is novel here is that I consider, for each terrestrial planet, degassing

II. DEGASSING

a mass of order 10^2 that of the existing planet—a loss of 98 to 99% of its initial mass. The perspective is qualitatively different.

If we look at the system as a chemical structure we are led to the concept of a pool or reservoir of atoms from which extracts can be drawn or constituents rejected. The initial pool is just a sample of the nebula. From this pool certain constituents are removed. The composition of the remaining pool is that of the new proto-planet. In Chapter 6 this proto-planetary pool will be taken as the initial pool for the discussion of the development of chemical structure through geological time.

A. General Considerations

In the Laplacian model of the evolution of the proto-solar system from a predominantly hydrogen-filled nebula, as presented here, we recognize two initial sizes of bodies.

1. $M/M_{(sun)} \gtrsim \alpha$, $\alpha \sim 0.1$, for H-fusion. Our system has only one such body, the Sun. The Jovian planets with $M/M_{sun)} \sim 10^{-3}$, although net emitters of radiation, do not have internal temperatures (typically $\sim 10^5$ K) sufficiently high for H-fusion ($\gtrsim 5 \times 10^6$ K).

2. $M/M_{(sun)} \lesssim \beta$, $\beta \sim 10^{-5}$, for H-escape. Apart from the Jovian planets, the mass proportion of hydrogen in the smaller bodies of the solar system is very small. The value of β must lie between that of the Earth, the largest non-H planet, and that of Uranus and Neptune, the smaller H planets, namely $\sim 10^{-5}$.

   Given that the nebula contained a mass proportion of hydrogen of order $10^2:1$, we are forced to the striking conclusion that the terrestrial planets have lost the great bulk of their mass. For example, the proto-Earth, for the above ratio, would have had an initial mass of 600×10^{24} kg and then lost 594×10^{24} kg of hydrogen and helium to leave a residuum of 6×10^{24} kg.

The masses of the proto-planets. In the scenario presented here all the proto-planets originate with a similar chemical constitution, that of the local part of the solar system nebula. The Jovian planets retain their chemical constitution but the terrestrial planets lose the bulk of their volatiles.

An estimate of the initial masses of the proto-terrestrial planets can then be made as follows. Assume that the solar system nebula had the composition of the present Sun, abundance data for which are shown in Table 1.1 (the accuracy of the data items is at best a few per cent). The rock substance of the present Earth is more than 99% by mass composed of O, Si, Al, Fe,

4. PARTURITION OF THE PROTO-TERRESTRIAL PLANETS

Mg, Ca, Na and K. For the present estimate we can ignore the constituents which make up the remaining 1%.

1. If only O, Si, Al, Fe, Mg, Ca, Na and K are retained, (a) the present mass M is proportional to the sum of the mass abundances of the retained constituents, 13,130, and (b) the original mass M_0 is 1,168,420 so that $M_0/M \approx 90$.
2. If the situation is as in (1) except that only enough oxygen is retained to form the standard "oxides", namely 2255 (so that of the original 9400 an amount 6145 must be lost), the present mass is proportional to 5990 so that $M_0/M \approx 195$.
3. If the situation is as in (2) but all the carbon is retained we find $M_0/M \approx 54$. But there is not enough oxygen for all the carbon. If then we retain only the carbon for which there is enough oxygen after satisfying the constituents as in (1), namely carbon atoms 2246 (with about 2000 to be lost), then the present mass is proportional to 14,226 so that $M_0/M \approx 82$.

Plainly the mass ratio $M_0/M \sim 10^2$. In the calculations of Chapter 4 I have taken arbitrarily the value $M_0/M = 70$.

Collapse ratios. As the proto-solar system collapses, individual proto-planets begin to emerge as distinct entities. The size of the zone from which an individual planet draws its mass will be restricted by the conflict for matter from other proto-planets. On the assumption that the planetary orbits have not changed substantially in solar system time, it is possible to make a very crude estimate of the original size of a proto-planetary disc. An estimate of the "collapse ratio" is obtained from the geometrical mean $\sigma = (xy)^{1/2}$ of the distances x and y for planet j such that

$$x/(r_{j+1} - r_j) = u/(1+u) \qquad u = (m_j/m_{j+1})^{1/2}$$
$$y/(r_j - r_{j-1}) = 1/(1+v) \qquad v = (m_{j-1}/m_j)^{1/2}$$

except that for Mercury the value is x and for Neptune the value is y. This crudely assumes that matter reaches a proto-planet from a region of orbital extent proportional to (mass)$^{1/2}$: x and y are the orbital collection radii determined by competition with the two adjacent bodies, x for collection exterior to and y for collection interior to the orbit of the proto-planet. The choice of a geometrical mean value is arbitrary. Tabular values are σ/a, where a is planetary radius. The range is obtained from the spread of x and y around σ. The values of x and y lie within a factor of 2 of σ, and except for Saturn, within a factor of about 1.4 of σ. For the Sun the minimal ratio is that of the radius of Neptune's orbit to that of the Sun's radius.

II. DEGASSING

Table 4.1. Collapse ratios

	Ratio	Range factor
(Sun)	$(>6.4 \times 10^3)$	–
Mercury	4.2×10^3	–
Venus	4.6×10^3	(1.43)
Earth	5.6×10^3	(1.65)
Mars	4.3×10^3	(1.32)
Jupiter	6.7×10^3	(1.22)
Saturn	7.9×10^3	(2.20)
Uranus	21.4×10^3	(1.38)
Neptune	34.9×10^3	–

These ratios are given in Table 4.1. They are typically of order 5×10^3 for the terrestrial planets and somewhat larger for the Jovian planets.

Naïve model. In order for a particle to escape from the gravity field of a body, a prime requirement is that it have sufficient kinetic energy, of amount at least equal to the magnitude of its gravitational potential energy near the surface, to carry it to a distance large compared to the size of the body; and, of course, it must not bang into another molecule en route.

1. For a disc of mass $M_d = mM_0$ and radius $R_d = rR_0$ this condition defines the minimal escape velocity V such that

$$V^2 = 2\sqrt{(2)}(GM_0/R_0)m/r$$

   where M_0 and R_0 are reference mass and radius. (Note that for a spherical body the relation is the same except that the factor $\sqrt{(2)}$ is dropped.)

2. The distribution of speeds of a molecule is determined by its temperature $T \equiv \theta T_0$ and atomic mass ratio μ. The (most) probable speed U is such that

$$U^2 = 2(kT_0/m_H)\theta/\mu$$

   where T_0 is a reference temperature.

3. Escape depends on the speed ratio $w \equiv V/U$. A particle will be unable to escape if $w \gg 1$. Let us naïvely assume that particles are trapped if $w \geq \xi$ and otherwise escape, say for $\xi = 10$. Thus particles are trapped for

$$\xi^2 \theta/\mu = \sqrt{(2)}\eta m/r \qquad \eta = (GM_0/R_0)/(kT_0/m_H)$$

The quantity η is a measure of the ratio of the gravitational potential energy

4. PARTURITION OF THE PROTO-TERRESTRIAL PLANETS

of a particle to its thermal energy. It is a key parameter in describing the escape process (see Öpik, 1963). Taking $M_0 = M_{(earth)}$, $R_0 = R_{(earth)}$ and $T_0 = 300\,\text{K}$ (a typical value for the terrestrial planets), we have $GM_0/R_0 = 63\,\text{MJ/kg}$ and $kT_0/m_H = 2.5\,\text{MJ/kg}$, the ratio of these two specific energies, gravitational to thermal, being about 25—so that for trapped particles

$$\theta/\mu \leqslant 0.36 m/r$$

This relationship is shown in Fig. 4.1. The outstanding feature is that the planets fall into two groups: the Jovians, for which all species now are trapped; and the terrestrials, for which escape of low atomic mass volatiles occurs now—hydrogen and helium for all, and most volatiles for the Moon. (Note: The present-day values are plotted at $m/\sqrt{(2)}$ corresponding to escape from a spherical body rather than a disc.)

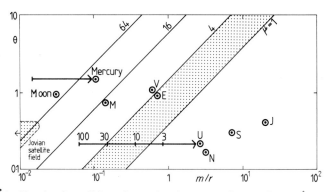

Fig. 4.1. Sketch of conditions for molecular escape from a disc. Temperature ratio θ as a function of mass/radius ratio m/r for various atomic mass ratios μ. Below the line particles are trapped; above they escape. The data for various bodies now is also plotted, with temperature ratio from current mean photosphere temperatures. The arrows refer to possible development paths.

Consider further the consequences of the development of the disc.

1. *Mercury*. The arrow shows the line of development for the example of an initial mass 100 times and an initial disc radius 1000 times, the present mass and radius. The ability of volatiles to escape is very high early in the development. This suggests that Mercury will be strongly depleted in low atomic mass elements.

    Similar comments apply to the other planets, the Moon being the most depleted, Earth the least.

II. DEGASSING

2. *Uranus*. The arrow shows a possible line of development—the numerical labels are points of initial disc radius ratio to the present radius for *no* mass loss. But for disc radius sufficiently large, some of the development occurs in the mass loss field. This indicates that, although now there is no escape from Uranus, there was during the early development of its disc.

   Similar comments apply to the other Jovian planets, Uranus being the most evolved, Jupiter the least.

This sketch has a number of obvious inadequacies. It merely presents a snapshot view and ignores the role of time. During its development, the disc changes both its physical and chemical structures. Thus there are powerful consequences of changes in m and r as the disc loses mass and as it contracts. Because the initial composition is largely hydrogen and helium there is a powerful sequencing of species discharge, so that the species are lost in order hydrogen, helium, ... and there is negligible discharge of a species until all those of lower atomic mass are completely lost. (Only for very small bodies, see later, is there any tendency for species to leave together.) These effects produce qualitative differences.

In a more detailed model, allowance also needs to be made for the molecular velocity distribution, as it affects the probability of escape; and some care is needed in determining the temperature and density of the level from which escape occurs. These matters are very complicated, indeed they cannot be dealt with in anything like a satisfactory manner; they are not, however, as potent ingredients of the model as those mentioned above.

B. The Energy Budget

The disc model of Chapter 3 retains all its features except for the presence of a mass flux. This requires a change in the budget, and the need to relate the mass flux to the given photosphere temperature. For the budget, whereas in the original disc model the drain on the energy stock is solely through the radiation loss, in this case the drain on the energy stock is dominated by that removed in the gas stream. There are thus two extreme disc models, one dominated by radiation loss and the other by direct potential energy loss. For the evaluation of the mass flux itself, a simple representation of the outer layers of the disc is used.

A mass flux f (kg/(m² s)) carries a power output, in the escaping stream of matter, of $2\pi R_d^2 \phi f$, where ϕ is the specific work function given by $\phi = R_d g_0 / \sqrt{(2)}$ with $g_0 = 2GM_d/R_d^2$, the gravitational acceleration near the disc surface. (The corresponding escape velocity V is such that $V^2 = 2\phi$.) Thus we have the following conservation relations.

For a mass and work function change M_1 to M_2 and W_1 to W_2 in time dt:

conservation of mass $\quad M_2 - M_1 = -2\pi R_d^2 f dt$

conservation of energy $\quad \frac{1}{2}(W_2 - W_1) = -2\pi R_d^2 \phi f dt$

where $W = \alpha_d G M_d^2 / R_d$.

In the units M_0 and R_0; taking a reference flux $f_* \equiv 1 \, \text{kg/(m}^2\text{s)}$; writing $\tau_0 = M_0 / R_0^2 f_*$; in dimensionless units

$$m_2 - m_1 = -2\pi r^2 f' dt' \qquad \frac{m_2^2}{r_2} - \frac{m_1^2}{r_1} = -4\sqrt{(2)}\pi m r f' dt'$$

For dt' sufficiently small, with $r \approx 1/2(r_1 + r_2)$ and $m \approx 1/2(m_1 + m_2)$, these two conservation relations—for given f', m_1 and r_1 and, say, given r_2—give m_2 and dt'. Thus by choosing a sequence of values of r_2 we can follow the corresponding mass evolution of the disc.

Note that the time scale is τ_0. For Earth values $\tau_0 = 4.67 \times 10^3$ year.

Once a central body is well established the gravity field will be enhanced near the central region of the disc. It has been unnecessary to take this into account in this model since for all the bodies of interest the bulk of degassing occurs early in the disc life before the emergence of the central body.

The mass flux. In this model we are not explicitly concerned about the details of mechanisms within the outer layers of the disc (envelope or atmosphere). The model requires only the temperature and density at the exo-base in order to evaluate the mass flux. To do this in a satisfactory and satisfying manner, merely using existing knowledge, would be a Herculean task. The simple approach taken here is inadequate in several ways. Its worst aspect is that it ignores the complex chemistry of the outer layers. For example, although Jupiter is a hydrogen–helium body, its outer layers contain ammonia and various hydrocarbons and distinct cloudy layers which have a profound effect on the local opacity coefficient and the thermal structure. A strictly phenomenological model is all that can be hoped for.

The mass flux is the total of the individual fluxes of the chemical species. For species i the mass flux $f_i = \xi_i \rho_i u_i$, where u_i is the probable molecular speed, ρ_i is the species density and ξ_i is the probability of escape; both ξ_i and u_i are functions dependent on the exo-base temperature and the species molecular mass. (For details see the Appendix. The model uses the Jean's escape mechanism.)

Role of model parameters. There are many parameters in this model. Some of them are known or lie within a known range (e.g. the molecular ratios and solar abundances); others affect the behaviour weakly (e.g. the

II. DEGASSING

choice of optical depth of the photosphere); and three have a dominant effect. These are the initial mass m_a and radius r_a of the disc and the photosphere temperature T_d. The disc structural parameters have their major influence through the escape velocity, and the photosphere temperature through the molecular velocities at the exo-base. In combination their effects are extreme.

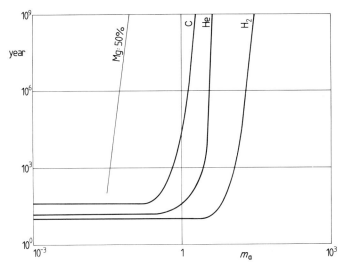

Fig. 4.2. Disc depletion time (year) as a function of initial disc mass ratio m_a. Lines are for complete depletion of the named species—to the left of a curve depletion is complete (other higher molecular mass species may also be depleted or partly depleted); to the right depletion is incomplete or negligible. Model parameters: $T_d = 200$ K, $r_a = 1000$.

The role of initial disc mass is illustrated in Fig. 4.2. For a model in which m_a is the only parameter which is changed this shows the time to complete depletion of a particular species.

Consider first the behaviour of H_2:
1. For $m_a \gtrsim 100$ there is no loss at all.
2. For $m_a \approx 30$ there is complete depletion at $t = 1$ Ga.
3. For $3 \lesssim m_a \lesssim 30$ there is a rapid increase of depletion time with mass; near the upper end $t \sim m_a^{18}$!
4. For $m_a \lesssim 3$ there is complete depletion at $t = 10$ year.

This is a very fierce function. It is like the behaviour of a switch. In effect there is a critical mass: for masses greater than critical, nothing special happens; for masses less than critical there is a profound change in the system, with a rapid loss of mass.

Consider next the behaviour of He. Keep in mind that in this model the He will be depleted only after all the H_2 is lost. The run of behaviour is similar to that of H_2 except that the critical mass is smaller by a factor of typically 5, and for a particular m_a the depletion time is much greater—e.g. in the ratio 10^3 at $m_a = 3$, and 10^9 at $m_a = 4$. This big change is merely with a change of molecular mass from 2 to 4.

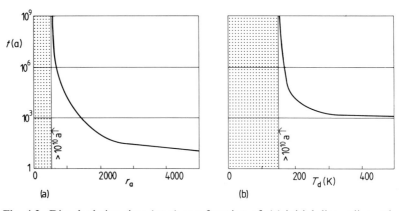

Fig. 4.3. Disc depletion time (year) as a function of: (a) initial disc radius ratio r_a with $T_d = 200$ K; (b) disc photosphere temperature T_d (K) with $r_a = 1000$. Model parameters otherwise as in Fig. 4.2.

This switch-like behaviour arises from the combination of the gravitational and molecular velocity effects modelled with the probability function ξ. On paper the function $(1 + w)e^{-w}$ (see Appendix) seems very ordinary. In reality its effects are potent and are at the heart of the presentation here, which sees all the planetary, satellite and minor bodies in a single perspective.

The role of r_a and T_d is shown in Fig. 4.3, which gives times for complete depletion of H_2 and He.

(a) When r_a is large, depletion is rapid. When r_a is sufficiently small, 500 in the case shown, depletion is negligible. This behaviour arises from the effect of r_a on the gravity field near the disc and hence on the escape velocity.

(b) When T_d is large, depletion is rapid. When T_d is sufficiently small, $\lesssim 150$ K in the case shown, depletion is negligible. This behaviour arises from the effect of T_d on the molecular velocity distribution.

II. DEGASSING

Table 4.2. Degassing model parameters for complete loss of H and He from proto-terrestrial planets.

	T_s (K)	r_a	r_d	$t(a)$
Mercury	450	1000	3.4	5×10^3
Venus	350	5000	36	7×10^5
Earth	300	5000	26	6×10^7
Moon	300	500	2.5	2×10^2
Mars	200	1000	6	4×10^5

C. Flushing the Proto-Terrestrial Discs

Model results for the proto-terrestrial planets, with the parameters of Table 4.2, are shown in Fig. 4.4. (In the model hydrogen is depleted as H_2.)

1. In contrast to the Jovian planets, which remain as hydrogen–helium bodies, the proto-terrestrial planets readily lose all their free hydrogen and helium.
2. In contrast to the very small bodies, discussed below, the depletion of rock substance constituents is negligible.
3. The time interval required for complete depletion of hydrogen and helium ranges from 2×10^2 year for the Moon to 6×10^7 year for the Earth.
4. For the more massive bodies (Earth, Venus and Mars) the depletion of hydrogen is almost complete before any helium loss. This is indicated by Y rising to about 0.8 at the moment $X = 0$.
5. For the smaller bodies (Mercury and the Moon) some helium loss occurs concurrently with hydrogen loss—especially so for the Moon.

Thus after an interval of at most 60 Ma the mini-discs of the proto-terrestrial planets are fully degassed and contracted to a size at which we might expect a central body to begin to grow, and possibly to have already begun to grow.

Simple degassing produces bodies enriched in Z-component material (for definition, see Chapter 1, Section V). If the degassing is only of hydrogen and helium the composition of the Z-component material will be unchanged from that in the nebula. Should degassing, however, be sufficiently intense for further species to be lost, the composition of the Z-component material itself would be altered.

For the terrestrial planets, this question arises in Chapter 6, where we find the strong suggestion that the terrestrial planets have different gross compositions—the Moon is depleted in FeO, Mercury is enriched. Furthermore, the Jovian moons, as largely ice bodies, are very different in composition from the terrestrial planets. The present Z-component material

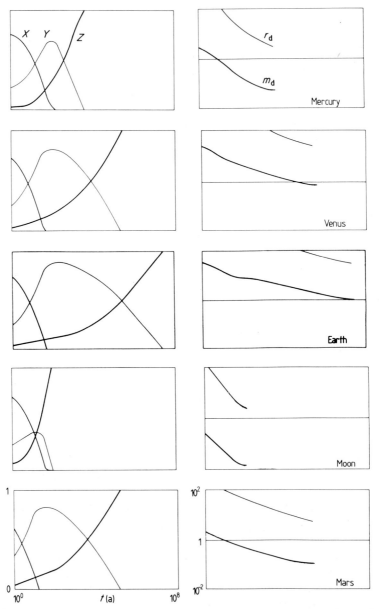

Fig. 4.4. Degassing of the proto-terrestrial planets. Mass concentration ratios X, Y and Z, and disc mass and radius, m_d and r_d, as functions of time t (year).

II. DEGASSING

is not the same everywhere. Some additional segregation mechanism must have operated.

Before proceeding, let us ask this question: Could (intense) degassing alone lead to the suggested enrichment? In principle the answer is yes. For a body of sufficiently small mass, stripping of species occurs up to high atomic mass ratios, so that, for example, the ratio MgO/FeO would be low for a small body and unaltered for a large body. In reality the answer is no. Even Mercury is too massive. Even for extreme choices of the model parameters, depletion of MgO is negligible. (In any event, if we restricted ourselves to this mechanism the Moon would also be depleted in MgO.)

D. Application to the Early Proto-Solar System

The early solar disc. This powerful segregation mechanism may have operated early in the life of the solar disc itself. If the disc were sufficiently cool, the loss of matter would have been negligible; if the disc were sufficiently hot, it would be dispersed before any central concentration could occur. These possibilities have been mentioned already in the qualitative discussion of Chapter 2. The escape model allows us to make some quantitative statements.

In order to focus our ideas, suppose that there has been substantial mass loss. Model data for a 90% mass loss (namely, original disc mass = $10M_{(sun)}$) are shown in Fig. 4.5. I have taken the initial nebula hydrogen fraction as $X = 0.976$, so the initial proto-Sun would have the estimated $X = 0.85$ (depleted now to about 0.8 by hydrogen fusion). Thus, for example, a disc of initial radius 10^4 AU, at photosphere temperature 50 K, degassing for 10 Ma, would have collapsed to a radius of about 1000 AU and thereafter be compositionally stable. This collapse rate would require central drift velocities of order 0.1 km/s. With an initial radius of 2×10^4 AU, depletion would be rapid and the final mass much less than $M_{(sun)}$. With an initial radius of 0.5×10^4 AU, the disc would lose a negligible mass and stay with a mass $10M_{(sun)}$.

The model therefore suggests severe limits on the physical and chemical structure of the early disc. Furthermore, it suggests that the original nebula may have been composed of low-Z ($\sim 10^{-3}$) material, as found today in "population II" stars. For the purpose of this book, however, it is sufficient to note that the mass and chemical composition of the solar disc was established after a prelude of perhaps 10 Ma when the disc had collapsed to a radius of order 10^3 AU.

Application to the proto-Jovian planets. When the disc mass is sufficiently large the effects of mass flux are muted or negligible. This is the field

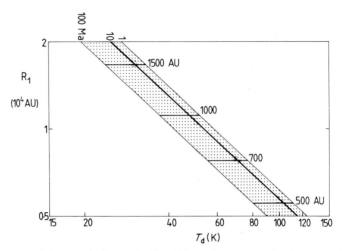

Fig. 4.5. Model example for early disc. Lines of constant development time as a function of initial disc size R_1, in AU, and photosphere temperature T_d (K). Development time for hydrogen mass fraction to fall to $X = 0.85$ from an assumed initial value of $X = 0.976$, corresponding to a 10 times hydrogen richer mixture than that of the proto-Sun. Disc radius at end of degassing indicated (500, 700, 1000, 1500 AU).

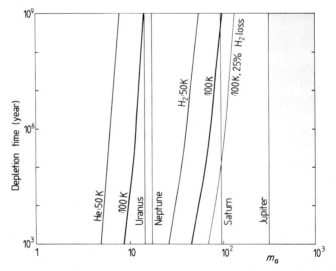

Fig. 4.6. Depletion time diagram for Jovian planets. Complete depletion of H_2 and He at $T_d = 50$ and 100 K, in year, as a function of initial disc mass ratio m_a. The line for 25% H_2 loss at 100 K is also shown. The planet lines are drawn for their present masses.

II. DEGASSING

of the Jovian planets. A summary of the circumstances is provided in Fig. 4.6.

1. *Jupiter.* With current mass ratio 318 and disc photosphere $T_s = 70$ K the mass loss is negligible. Relaxing the parameters somewhat we have the following results:
    (a) At $T_s = 70$ K, $r_a = 5000$, with extreme $m_a = 330$ (extreme = highest m_a to allow final mass equal to current mass), a mass loss of hydrogen as H_2 of 1% would occur in 1 Ga.
    (b) At $T_s = 100$ K, $r_a = 5000$, with extreme $m_a = 380$, a mass loss of hydrogen as monatomic H of 4% would occur in 1 Ga.
    (c) At $T_s = 100$ K, $r_a = 10,000$, with extreme $m_a = 500$, a mass loss of hydrogen as monatomic H of 15%, from $X = 0.80$ to 0.68, would occur in 0.4 Ga.
    Jupiter has a mass well above the critical for hydrogen depletion. At most there could have been perhaps a 20% loss. It is a truly primitive body.
2. *Saturn.* With current mass ratio 95 and disc photosphere $T_s = 50$ K, a noticeable mass loss is possible at $T_s = 50$ K, $r_a = 5000$; $m_a = 110$ gives the final mass equal to the current mass in 0.77 Ga with lost matter as H_2; as monatomic H, $m_a = 150$ requires 3 Ma; $m_a = 170$ requires 0.2 Ga; $m_a = 180$ requires 3 Ga. Thus at most the mass loss is about 50% (the original mass being twice the present mass).
3. *Uranus.* With current mass ratio 14.5 and disc photosphere temperature $T_s = 40$ K, substantial hydrogen mass loss can occur. Helium loss is negligible. At $r_a = 5000$ and $m_a = 55$, the hydrogen fraction, depleted as H_2, would fall in 0.2 Ga to give $X = 0.24$, $Y = 0.68$ and $Z = 0.08$. For loss as monatomic H an initial mass $m_a = 70$ would be completely depleted in H in 1 Ma. At the other extreme, a mass of $100 m_a$ would take 50 Ga for substantial hydrogen depletion with loss as monatomic H. For loss as H_2, an initial mass of five times the present mass is quite possible.
4. *Neptune.* This is quantitatively similar in behaviour to Uranus.

In summary, this point of view strongly suggests the following. Jupiter and Saturn are the hydrogen planets: at most they have lost 20 and 50% respectively of their original hydrogen, and are truly primitive bodies. Uranus and Neptune are the helium planets: they have lost a high proportion of their hydrogen and none of their helium, are enriched by perhaps a factor of 5 in their Z-component, but remain as rather primitive bodies.

Application to small bodies. The terrestrial proto-planets are sufficiently massive to retain the bulk of the original Z-component material. For

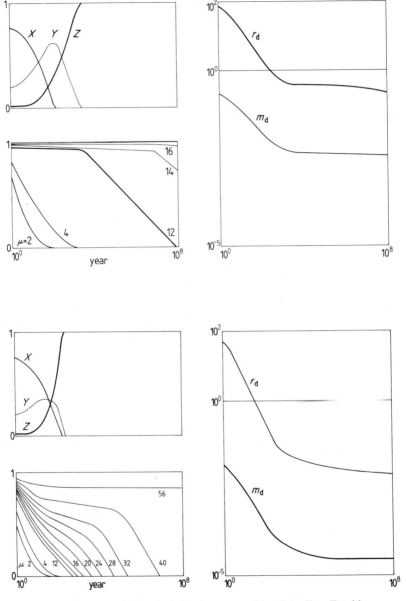

Fig. 4.7. Degassing and chemical development of small bodies. Freckle mass, m_d and radius r_d; mass concentration ratios, X, Y, Z; and mass concentration for species of atomic mass, 2, 4, 12, 14, 16, 20, 23, 24, 27, 28, 32, 40, 56 (H_2, He, C, N, O, Ne, Na, Mg, Al, Si, S, A and Ca, Fe). (a) $m_a = 0.3$; (b) $m_a = 0.03$. With $r_a = 100$, $T_d = 400$ K.

II. DEGASSING

somewhat smaller bodies, however, depletion of Z-component material becomes pronounced. This is illustrated for some model bodies in Fig. 4.7.

1. $m_a = 1$ (not shown). Apart from the higher mass this model body has the same parameters as those of this example. Yet it has negligible loss of any Z-component material.
2. $m_a = 0.3$. Some He departs with the H. Depletion of $\mu = 12$ material, carbon, becomes strong at the end of He depletion and is complete in 100 Ma. Depletion of the other constituents is minor during this interval.
3. $m_a = 0.03$. Depletion of He is concurrent with depletion of H. Depletion of all constituents except for $\mu = 56$, Fe, is extreme. After 2 Ma the body mass is 3×10^{-5} ($= 1.8 \times 10^{20}$ kg), composed of 94% Fe and 6% Ca. Though small compared to the planets, this residual object—a small Fe-dust disc—would be one of the larger small bodies.

It is helpful to be able to refer to the parts of the structure in a simple, direct manner. I use the word "disc" to refer to any thin patch of gas, specifically the disc of the proto-solar system—and although modelled as if it were of circular plan the shape may be irregular. The term "mini-disc" is used for a sub-solar system patch destined to become a proto-planet with its own satellite system. For even smaller patches the term "micro-disc" is a possible choice—I use the word "freckle". This implies a small transient patch which grows and evaporates itself in a swarm of freckles which continually interchange matter with one another.

This part of the system behaves in a different manner from that of the system as a whole. Broadly the gross system behaviour is a one-shot process, a simple monotonic passage through the system structure space, as in the sequence a-b-c-d-etc.; the freckle field is a multipass system, as in the sequence a-b-a-c-d-b-e-etc.—an elaborate regurgitation. If, in addition, the freckle field is in a sufficiently non-uniform gravitational field, as it would be near the proto-Sun, there is the opportunity for a weak central drift of heavier components left naked towards the end of a freckle's life. Here, then, is a mechanism for segregation of the non-volatile components. I do not explore this topic further here—but I see it as a very fruitful avenue of investigation.

A body of sufficiently small mass, at a given distance from the proto-Sun, will be completely dispersed. It is perhaps no surprise that there are no minor planets near the Sun. The nearest planet, Mercury, is indeed the smallest and least massive planet—it is about half the size and mass of Mars, the next in size—but the existence of one planet does not make a "summer".

The segregation processes occur in the presence of abundant hydrogen. This is particularly so for small bodies for which the mass flux of all the constituents is significant—large bodies lose their chemical constituents sequentially and as a consequence have only minor hydrogen present during their later development. In these "reducing" conditions, the occurrence of free metals can be expected. Thus if a body, sufficiently small, is almost completely dispersed an Fe-rich residuum remains.

Segregation and dispersal of freckles in the gravity field of a nearby large body will be enhanced. The more volatile material will be moving more rapidly than the residual material. It will have a greater opportunity to recombine into new freckles and will have less tendency to drift towards the central body. The residual material, however, especially if it is in solid lumps which are less likely to be vaporized again, will have less mobility of its own and will therefore have an enhanced tendency to drift towards the central body. Thus we would expect a depletion of higher atomic mass in the outer portions provided there is a large central body. Thus we would expect Mercury to be enriched in FeO, the Moon to be depleted in FeO and the Jovian moons to be depleted in FeO.

The occurrence of the numerous small moons of the Jovian planets is most remarkable. The escape velocity from these bodies is very small and in isolation we would expect all volatiles to have very easily escaped during their formation. Plainly this is not so. Clearly the role of the gravity field of the massive central body is important. None of the volatiles can escape from the vicinity of the central body. Thus we envisage in the outer envelope of the central body an accumulation of volatiles which remain available to accumulate into separate moons.

This suggests also that those asteroids which are icy formed in the vicinity of one of the Jovian planets.

In Chapter 3 we saw a system segregating itself in its own gravitational field. That system was taken to be chemically homogeneous. In this chapter we see how chemical inhomogeneities develop. The system has reached a stage of development dominated by chemical segregation—still under the control of the system's own gravity field. This stage lasts a few hundred million years. At its close, perhaps about 4.5 Ga ago, the segregation of hydrogen and helium is complete.

III. CORE FORMATION

There are many possible paths between the state of the original gaseous nebula and the rock substance terrestrial planets. One such path is presented

III. CORE FORMATION

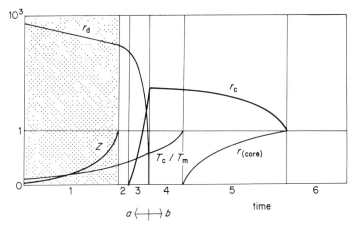

Fig. 4.8. Degassing and core formation scenario. Properties as a function of time, stages 1–6 referred to in text, 1–3 in first part of this chapter, 4–6 in second part of this chapter. Radii of disc, r_d; central body, r_c; molten core, $r_{(core)}$; composition, Z; central temperature ratio, T_c/T_m. Degassing stage shaded. Axes not linear and not to scale—but curves taken from model of Mars. Total time interval shown about 200 Ma.

here. For the proto-terrestrial bodies the final stages of the hydrogen era are soon over. In this prelude to geological time two key events occur: the loss of volatiles to give the residual rock substance material itself—already described; and the accumulation and melting of that material to produce completely liquid bodies, the proto-terrestrial planets in their naked form, just before the onset of geological time.

Let us therefore now look at the proto-planetary object not from the point of view of the volatiles but with emphasis on the Z-component material which will remain. We have the following sequence of events, illustrated in Fig. 4.8.

1. Loss of volatiles occurs from a homogeneous gaseous body. Degassing goes to completion. The body is still large. Initially temperatures will be low so that most of the Z-component will be as oxides, possibly in temporary aggregates. The matter behaves as a slightly dusty gas.
2. The disc continues to contract.
3. Ultimately a central body begins to form, taking up matter from the surrounding disc, until the central body reaches its maximum size. The disc is exhausted. The remaining object is a nearly spherical body also composed entirely of a cool dusty gas.
4. The dusty gas sphere begins to contract. Temperatures rise, until in a central region Z-component phases can melt.

70   4. PARTURITION OF THE PROTO-TERRESTRIAL PLANETS

5. Two more or less distinct regions appear. These molten phases will at first be sparsely scattered and the central region will be a zone of droplets embedded in a turbulent gas. This could be called the "thunder cloud" stage. With further collapse, droplets temporarily coalesce in the central region to produce a foam. A dusty envelope surrounds a distinct foam core. The core becomes progressively a more or less homogeneous liquid body with an outer foaming layer.
6. A liquid planet surrounded by a thin atmosphere of minor residual components remains.

Stages (1)–(3) have already been discussed in the first part of this chapter. Now we concentrate attention on the core formation stages (4)–(6).

A. Emergence of the Proto-Terrestrial Central Bodies

Soon the contracting mini-discs are largely depleted of their hydrogen and helium. They continue to contract. They are now composed of a cool dust,

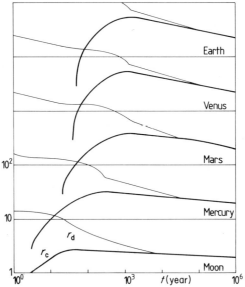

Fig. 4.9. Proto-terrestrial planets, emergence of the central body from the Z-component disc. Radius ratio of the disc and central body as a function of time t (year). The axes labels refer to the bottom curves for the Moon—the other curves are displaced successively upwards by an order of magnitude. (Model parameters: disc photosphere temperature = 150 K for Mercury, Venus, Earth and the Moon, and 100 K for Mars; $r_1 = 200$; $r_2 = 50$; $r_3 = 10$; and r_* is proportional to planet current radius with $r_* = 30$ for Earth.)

III. CORE FORMATION

of simple oxides or perhaps simple silicate assemblages, diluted by residual hydrogen and helium. The role of the mass flux is now negligible; the rate of contraction is controlled by the radiation power loss. The mini-discs now behave just like the mini-discs of the proto-Jovian planets (and the disc of the proto-Sun) except that the matter has a higher mean atomic mass.

As the mini-discs contract a central body ultimately begins to grow. The growth of this central body is illustrated in Fig. 4.9 for the proto-terrestrial planets.

There are no data with which this stage of the model can be calibrated. This stage of the model is therefore strictly hypothetical.

For a model Earth the central body reaches its maximum size, assumed to be $30R_{(earth)}$, at 1.5 ka after the termination of degassing; the disc is exhausted, at assumed central body ratio of $10R_{(earth)}$ in 80 ka. This is a short stage for all the proto-terrestrial planets—all are isolated spherical bodies within 10^5 year.

B. The Energy Budget

The proto-planet continues to collapse and its interior heats up from the released gravitational energy. If the body is massive enough, melting will commence in the central part of the body and a liquid core will begin to form.

The model is the same as that for degassing with the following modifications.

1. Apart from the presence of small amounts of volatiles it is for an entirely Z-body of fixed mass.
2. There may be a liquid core for which allowance must be made. In particular, the gravitational energy term will involve a contribution from both the envelope and the core.
3. If there is a liquid core the gaseous phase will be confined to an outer envelope.

Consider a spherical body with an outer envelope of dusty gas and a growing core of molten rock substance. Let the total mass be M_0, the mass of the envelope M_g, the mass of the core M_ℓ, the envelope radius R_g and the core radius R_ℓ. It is convenient to take a reference density ρ_0, a reference radius R_0 such that $M_0 = \frac{4}{3}\pi \rho_0 R_0^3$, and write $M_\ell = m_\ell M_0$, $R_\ell = r_\ell R_0$ etc., defining the dimensionless mass m_ℓ, radius r_ℓ etc.

In the numerical model, as reported here, I have taken the core as a homogeneous body of uniform density ρ_0. This is a good approximation when the core mass is small. For the larger proto-planets Earth and Venus, however, when the core mass approaches that of the entire body it is ordinarily a poor approximation; since compressibility is ignored the core radius will be overestimated by up to 20%, but this is of little consequence in this crude model. With this simplification, conservation of mass requires

$$m_\ell = r_\ell^3 \qquad m_g + m_\ell = 1$$

The envelope is modelled as a polytropic gas shell of mass $m_g = 1 - m_\ell$, inner radius r_ℓ and outer radius r_g, to be determined from the application of the boundary conditions. These are nominally $\rho = 0$ at $r = r_g$, and appropriate conditions at the interface $r = r_\ell$. Clearly conditions at the envelope–core interface, $r = r_\ell$, will be complex. I make the simplifying assumption that the melting temperature T_m is reached at the interface and that T_m is constant throughout the entire process of core formation. I have taken the core temperature as uniform and ignored the change of melting point as a function of pressure. The evaluation of r_g is straightforward (for details see the Appendix). For the given m_g and an assumed r_g the equations for the polytropic envelope give a value for the envelope temperature T_ℓ at $r = r_\ell$; r_g is then successively adjusted until $T_\ell = T_m$.

The development of the body is controlled by the loss of energy as radiation drawing on the gravitational energy resource of the body. The total work function W arises from the contributions of the envelope and core, so that with reference work function $W_0 = GM_0^2/R_0$, the dimensionless work function

$$w \equiv \frac{W}{W_0} = \alpha_\ell \frac{m_\ell^2}{r_\ell} + \alpha_g \frac{m_g^2}{r_g}$$

where the shape factors $\alpha_\ell = 0.6$, $0.6 \leqslant \alpha_g \leqslant 0.86$ ($\alpha_g \approx 0.6$ when the envelope is thin and $\alpha_g \approx 0.86$ when thick). Taking the temperature scale as T_m and the luminosity scale $L_0 \equiv 4\pi R_0^2 \sigma T_m^4$ so that the dimensionless luminosity $l = L/L_0$, and measuring time in units $\tau_0 \equiv W_0/L_0$, conservation of energy requires

$$\tfrac{1}{2}\Delta w = l \Delta t$$

The luminosity L is the net radiation loss from the interior of the envelope. It is determined by the temperature, density and opacity of the envelope (for details see the Appendix).

C. Growth of the Cores of the Proto-Terrestrial Planets

As the proto-planet collapses, one of two distinct possibilities occurs. The temperature of the central region rises and it may rise sufficiently high to melt or remelt any matter falling into it. For bodies of sufficiently small

III. CORE FORMATION

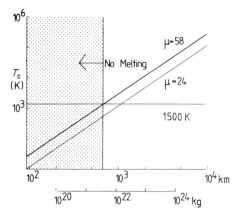

Fig. 4.10. Core-formation model. Central temperature T_c (K) as a function of gas sphere radius (km)—a mass (kg) scale is also shown for material of constant molecular mass μ.

mass this will not happen and a core of molten or partially molten material cannot form.

The mass of the smallest body which can form a core is found as follows. Consider the proto-planet as a polytropic body of given mean molecular weight μ, and assume its central temperature $T_c > T_m$, where T_m is a given melting temperature. A particular case is shown in Fig. 4.10. For example, for $\mu = 58$ (an oxide gas) and $T_m = 1500$ K no central melting occurs for bodies of mass less than about 4×10^{21} kg (and radius 700 km).

Very small bodies which have formed directly from the solar disc will not have passed through a molten core phase. Where there is, however, obvious evidence of melting in small bodies found today they must have been produced as fragments of larger objects. Large bodies such as the terrestrial planets and the Moon will certainly have been through a molten core phase.

Core onset. For bodies smaller than a critical mass no core is possible. For larger bodies a core forms but not immediately. Development tracks are illustrated in Fig. 4.11.

1. For r_g sufficiently large the body is entirely a dusty gas with no core. This stage is protracted for bodies of small mass.
2. At a critical value of r_g a core begins to form. With, for example, $T_m = 2500$ K the critical values of the radius ratio of the gas sphere (and the corresponding mean densities in kg/m³) are: Mercury 13.2 (4900); Venus 78.3 (4725); Earth 91.8 (5000); the Moon 4.25 (3000); and Mars 18.3 (3550).
3. Thereafter the core mass grows and r_g continues to decrease.

4. PARTURITION OF THE PROTO-TERRESTRIAL PLANETS

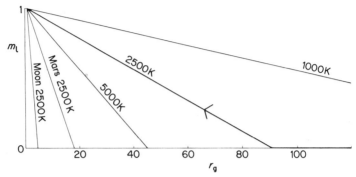

Fig. 4.11. Core formation. Core, envelope path with $\mu = 58$ for constant interface temperature: Earth 1000, 2500 and 5000 K; Mars 2500 K; the Moon 2500 K. Core mass ratio m_ℓ as a function of gas sphere radius ratio r_g.

Structure during core formation. The internal structure during core formation is illustrated in Fig. 4.12, which gives the radial temperature distribution for various gas envelope radii.

1. While the core is small the temperature structure is nearly indistinguishable from that for a wholly gaseous body.
2. Once the core size approaches that of the final body and the mass in the envelope is small the temperature profile outside the core becomes progressively distinct from that of a wholly gaseous body. The thermal interface at the liquid core–gas envelope boundary becomes sharper until the liquid core is surrounded by a thin "atmospheric" shell.

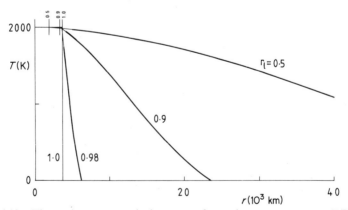

Fig. 4.12. Thermal structure during core formation: temperature $T(K)$ as a function of radius for indicated values of core radius ratio $r_\ell = 0.50$, 0.90, 0.98 and corresponding $r_g = 21.1$, 6.4, 1.7. The interface position is indicated by the short vertical lines. Model data for proto-Mars with: nominal $T_m = 2000$ K; $\mu = 58$; $\rho_0 = 3300$ kg/m^3; photosphere temperature nominally zero.

IV. CORE MASS

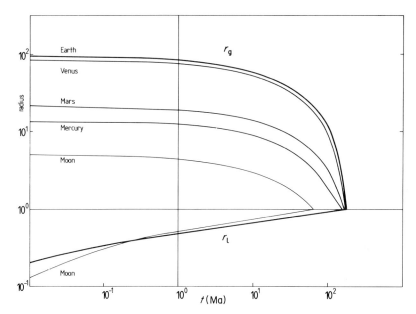

Fig. 4.13. Illustration of the growth of the liquid cores of the proto-terrestrial planets. Radius of the dusty gas envelope r_g and radius of core r_ℓ as a function of time t (Ma) from the moment of onset of the core. Except for the Moon, the r_ℓ curves are nearly indistinguishable. (Model parameters: reference density = 3300 kg/m^3; mean atomic mass = 58; envelope–core interface temperature = 2000 K.)

Core mass as a function of time. The growth of the cores for model terrestrial planets is shown in Fig. 4.13.

1. Once melting starts the early growth of the cores is rapid. The early response of the envelope contraction is slow.
2. The final growth of the cores is slow and the envelope response rapid.
3. The form of the development for all the terrestrial bodies is remarkably similar, except that the Moon's envelope is exhausted early. After a time interval of less than 200 Ma all the proto-terrestrial planets have collapsed their envelopes. Their parturition is complete.

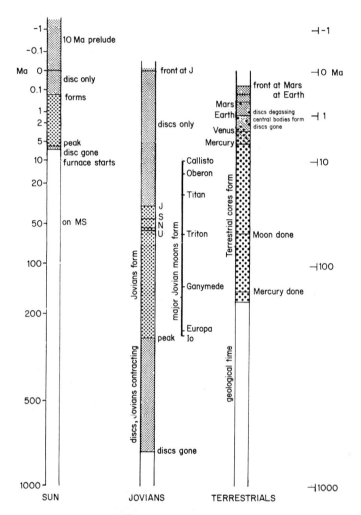

Fig. 4.14. Early solar system chronology. Schematic summary of the results of Part I for: the Sun; the Jovian planets; the major moons of the Jovian planets; the terrestrial planets. The distinct stages shown as a function of time, in Ma, are: the discs, birth and death (light shading); central bodies, birth and peak size (medium shading); terrestrial cores, birth and complete growth (heavy shading); initial emergence of the moons of the Jovian planets from their planets' discs. For the individual terrestrial planets there is overlapping in time of the individual stages of: discs degassing; growth of the central bodies; termination of the discs; and core birth, during 0.5 to 5 Ma—only the emergence of the central bodies from their individual discs is indicated. (The scale is proportional to the cube root of time, with zero time at the moment the solar disc front passes the orbit of Jupiter. The time values have only a qualitative significance.)

IV. CHRONOLOGY

It is helpful when considering one body to be aware of what the others are doing. We have little more than guesswork to help string together the sequence of events into a diary but, using the scenario of Part I of this book, I have constructed the chronology shown in Fig. 4.14. The reader may find it of value to try to construct an alternative. The major events occupy 200 Ma, by which time the Sun is well established on the main sequence; the planets and their satellites exist, although a small mass fraction remains in the discs of the Jovian planets; the structure of the Sun and Jovian planets is established for the remainder of solar system time; and the terrestrial planets have been born as molten bodies.

For the proto-terrestrial planets the hydrogen era is over. Although some hydrogen remains as a constituent of water substance, most of the hydrogen and helium has been lost. Some minor volatiles may remain to provide a proto-atmosphere. Soon the surface of the liquid body will cool sufficiently to allow new solid phases to occur. Geological time is about to commence.

PART II
THE GEOLOGICAL ERA

Early in solar system time, within an interval of order 100 Ma, the proto-terrestrial planets and other minor objects have emerged as hot bodies, the larger ones entirely molten, surrounded by thin residual envelopes of volatile material, the proto-atmospheres. Each body cools, low density, low melting point rock substance material accumulates in a near-surface zone, freezing commences, local geological time has begun. Further cooling, controlled by the proto-atmosphere and the proto-crust, allows the onset of a succession of liquid and solid phases of volatile material on the surface.

Chemical fractionation within the body interior progresses, rapidly during the first 1 Ga, to build the chemical and physical structures found today. The release of gravitational and thermal energy controlled by interior convection sets the rates of these processes. The structures can be calibrated with data from surface and astronomical measurements and particularly from data from upper mantle rocks. Information on planetary volcanism provides strong constraints on our ideas of the development of the structures of the terrestrial planets.

Message from the upper mantle of the Earth.

A thin slice, area shown 1 × 1 cm, of a xenolith from the Wesselton Kimberlite pipe, South Africa.

A garnet-peridotite, predominantly of large magnesium-rich olivine crystals, with two pyroxenes and minor magnesium-rich garnet, partly serpentinized. The mineral assemblage indicates that the xenolith originated from a depth of at least 75 km. (Author's collection.)

CHAPTER 5

Onset of Geological Time

I. INTRODUCTION

After a parturition interval which may last for 100 Ma for a particular terrestrial planet, it emerges as a new object. The stage of divesting itself of its "placenta" of hydrogen, helium and the bulk of its minor volatiles is over. Mass loss ceases. A fixed amount of matter remains—a liquid interior and a possible residual proto-atmosphere. Henceforth the body develops solely by rearranging itself.

There are many possible paths through the phase space of the body—paths characterized by the onset and disappearance of distinct phases, a process of progressive segregation of matter, from a birth as a vigorous hot fluid body till the senility of a fully solid cool body which is geologically dead.

In the scenario presented here all the terrestrial planets enter their geological phase as fully liquid bodies of molten rock substance. All but four of them (Venus, Earth, Mars and Titan) are essentially naked, without a substantial proto-atmosphere. For the moment I wish to concentrate attention on those bodies which have been veiled by dense and opaque proto-atmospheres.

This dense atmosphere acts as a strong barrier to energy transfer and provides through its interaction with the surface layers the essential controls on the early geological development of the planet. In the model these controls are represented by two parameters: ξ, a measure of the transmissivity of the

82   5. ONSET OF GEOLOGICAL TIME

atmosphere; and δ, a measure of the thickness of the near-surface zone of the initially liquid planet.

There are two essential stages in the early geological development: suitable crustal material needs to be collected in a surface zone and then the initial permanent crust can form.

The discussion refers successively to the proto-geological stage of: Earth, in which the ideas are developed, and there is a single dominant phase change, namely the deposition of largely liquid water; Venus, in which no phase change occurs in the atmosphere; Titan, to be compared with Earth except that, instead of liquid water, there is liquid methane; and Mars, the most elaborate, with three dominant phase changes, deposition of liquid water and its subsequent freezing and the deposition of solid carbon dioxide.

II. PRE-CRUSTAL FRACTIONATION

While the envelope is thick and opaque, the rate of loss of energy from the interface (the surface) between the envelope and the liquid interior is small: interior convection will be sufficiently vigorous to maintain a well-mixed interior, and the temperature of the interior will be close to the MPD (melting point as a function of depth) of the interior material. As the envelope clears, the rate of loss of energy through the envelope rises: the surface temperature begins to fall, and the first stage of global fractionation can begin.

This first stage of fractionation is envisaged as proceeding as follows. From the outset, impermanent solid material can form at the surface—this will be a high melting point fraction, initially not very different in composition from that of the deep interior. Slabs of this dense material will founder and fall into the interior, to be remelted. There will be vigorous stirring of the near-surface zone from convection of the liquid, and by the falling slabs. Also some crystals of low density may form at intermediate depths: they will be small and readily circulated in the convecting layer, growing as they are transported upwards, remelting as they are transported downwards. Progressively those constituents which give lower density material will accumulate preferentially near the surface to produce a near-surface zone of fractionated material of low density. This material is of low melting point.

At a later time the near-surface material will have a MPD close to that of a low melting point extract of the original interior material. Throughout this time the entire interior, including the near-surface zone, will be almost completely liquid. Once the surface temperature falls to that of the MPD of the fractionated material the first permanent crust can form: the pre-crustal stage is over.

Here we are interested in how long the pre-crustal stage takes. Let us make a simple model. Consider the idealized situation sketched in Fig. 5.1.

II. PRE-CRUSTAL FRACTIONATION

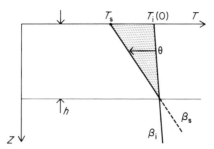

Fig. 5.1. Fractionation schema. Temperature profile near the free liquid surface.

A zone of depth h has been cooled. This zone is bounded by the lines: $T_i = T_{i0} + \beta_i z$, that of the MPD for interior material; and $T = T_s + \beta_s z$, the MPD for the surface zone material. Anticipating the discussion of Chapter 8, $\beta_i \approx 10^{-3}$ K/m and $\beta_s \approx 4 \times 10^{-3}$ K/m. As the surface temperature T_s falls, the size of the zone increases—until T_s reaches the liquidus temperature T_ℓ of the low melting point fraction.

In reality, all of this fractionated layer cannot be composed entirely of the low melting point fraction. The effect of the mixing in the fractionation process will be to produce a zone in which the concentration of the low melting point fraction is high near the surface and falls to zero in the interior below the fractionated zone. As a consequence, the difference of melting point gradients, $\Delta\beta = \beta_s - \beta_i$, will not be constant as assumed in the model.

The duration of the process will be controlled by the necessity to remove the thermal energy from this zone. Thus the total energy loss per unit surface area

$$\varepsilon \approx \rho c \bar{\theta} h = \frac{\rho c}{2\Delta\beta}(\Delta T)^2 \quad \Delta T = T_{i0} - T_s, \; \Delta\beta = \beta_s - \beta_i$$

with $h \approx \Delta T/\Delta\beta$.

Conservation of energy requires the surface heat flux (Section III) $f_s = d\varepsilon/dt$, where $f_s = \xi_0 \sigma T_s^4$, and we take $\xi = \xi_0$, a constant, thereby assuming no substantial change in the radiation transfer properties of the atmosphere during this stage. Hence

$$\frac{dT_s}{dt} = -\frac{\Delta\beta \xi_0 \sigma}{\rho c} \frac{T_s^4}{(T_{i0} - T_s)}$$

Writing

$$y = T_s/T_{i0}$$
$$\tau_0 = 5\rho c/6\Delta\beta \xi_0 \sigma T_{i0}^2$$
$$t/\tau_0 = \phi(y) = (y^3 - 3y + 2)/5y^3$$

which satisfies $y = 0$ at $t = 0$ and gives $y = \frac{1}{2}$ at $t = \tau_0$. The form of this function is shown in Fig. 5.2.

The process (with a fully fluid zone) ceases when $T_s = T_\ell$. In practice this will be for values near $y \approx \frac{1}{2}$. For example, with $\Delta\beta = 10^{-3}$ K/km, $T_{i0} = 1650$ K and $\xi_0 = 10^{-3}$, we have $\tau_0 = 5 \times 10^5$ year, after which time the surface temperature would have fallen to half its initial value.

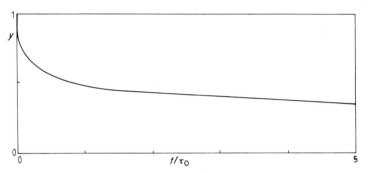

Fig. 5.2. Fractionation function, surface temperature ratio $y(t/\tau_0)$.

For the values used here with $T_\ell = 1000$ K, $y_\ell = 0.6$ and $\Delta\beta = 3 \times 10^{-3}$ K/m, the first permanent crust will form after a fractionation interval of about 6×10^4 year with $h \approx 150$ km.

For a planet such as Mercury, which after the H-degassing stage has an insignificant proto-atmosphere, so that $\xi_0 \approx 1$, the time scale of the fractionation stage is of order 10^3 year. Compared to the degassing interval this is very short. Hence the fractionation stage would be contemporaneous with the final part of the degassing.

The Surface Slag

Near the surface this picture is oversimplified. We need to envisage a near-surface zone in which slabs of material are continually being formed as rafts of surface material and then foundering. As they sink they will stir a layer of fluid and bring up hotter fluid from depth.

We are interested in the surface zone in the first instance in order to determine the constraints on the contribution to the atmosphere–surface zone energy balance. There are several possibilities.

If for example the planet were fully liquid, there would be immediately below the surface a thermal boundary layer of thickness δ determined by the interior convection, namely $\delta \approx 9(\kappa v/\gamma g \theta)^{1/3}$ (see Chapter 8), where v is the kinematic viscosity of the liquid and $\theta = T_i - T_s$. The values of δ are typically less than a few metres. For example, for a "granitic" liquid, δ ranges between 5 and 0.1 m, over T_i between 1000 and 1600 K.

It is most unlikely, however, that the surface of the planet is "clean" and mirror-smooth. The surface will likely be covered by a layer of more or less solid material—I shall refer to it as "slag"—which is continually being broken up and foundering to be remelted in the interior. Individual pieces of this slag will probably be porous because of vesiculation from partially trapped volatiles—and may actually float as a pumice-like slag.

The thickness of this slaggy zone will be taken as a phenomenological parameter of the model described below. Observation of lava lakes suggests a thickness of order $10-10^2$ m.

III. SIMPLE RADIATION MODEL OF THE PROTO-ATMOSPHERE–CRUST

In describing the development of the crust the key quantity is the temperature of the interface between the proto-atmosphere and the liquid/solid planet. The surface temperature is obtained through the balance of energy inputs and outputs to the atmosphere and inputs from the interior (a negligible contributor to the energy balance today for the Earth but a strong input during proto-geological time).

The solar flux at orbital distance r is $f = L_{(\text{sun})}/4\pi r^2$. The mean solar input flux $l = f/4 = \sigma T_*^4$, where T_* is a mean temperature. For a rapidly rotating black body, T_* is the mean surface (body) temperature. For a body with an atmosphere this temperature will be reached at about one optical depth into the atmosphere, the planet's photosphere—if the atmosphere is thick, usually near the top of the cloud layer. These temperatures (K) and the working values used in the model studies are: Mercury 447.6 (450); Venus 328.1 (330); Earth 278.6 (280); Mars 226.3 (225); Jupiter 122.3 (120); Saturn 90.5 (90);

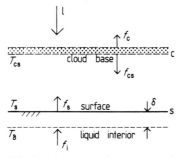

Fig. 5.3. Model atmosphere–crust schema.

Uranus 63.6 (65); and Neptune 50.2 (50). Satellite temperatures are taken as those of the central planet.

Consider the following model sketched in Fig. 5.3.

The atmosphere is represented as a layer of net transmissivity ξ such that $0 < \xi < 1$ topped by a "cloud" layer. The rock substance surface zone has a cooled region of thickness δ. For the system in quasi-equilibrium:

$$f_s = f_i + f_{cs} \qquad f_c + f_{cs} = f_s + I$$

where $f_c = \sigma T_{cs}^4$, assuming negligible opacity above the cloud tops; $f_{cs} = \xi \sigma T_{cs}^4$, the net flux from the cloudbase; $f_s = \xi \sigma T_{cs}^4$, the net flux from the surface; and $f_i = K(T_\delta - T_s)/\delta$, the input from the interior, where δ will be determined by conditions in the interior. Hence we have

$$\left(\frac{T_s}{T_*}\right)^4 = 1 + \chi\left(\frac{T_\delta - T_s}{T_*}\right) \qquad \chi = \frac{(1+\xi)K}{\xi \delta \sigma T_*^3}$$

Given (ξ, δ), this relation is readily solved (iteratively) for T_s.

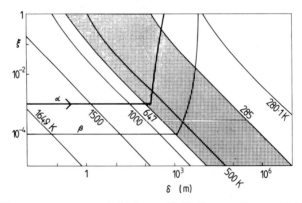

Fig. 5.4. Transparency, crustal thickness (ξ, δ) diagram for model Earth proto-atmosphere–crust. Lines of constant temperature and two development paths α, β.

Note that we are only interested in the case $T_s < T_\delta$. Furthermore, in the model presented here there is a built-in discontinuity, namely, for $T_s > T_m$, the temperature of the granitic liquidus, set $T_\delta = T_i$, the deep interior temperature; for $T_s < T_m$ set $T_\delta = T_m$.

This model does not pretend to be anything other than a simple phenomenological model described by two parameters.

Model behaviour is shown in Fig. 5.4, which shows lines of $T_s =$ constant (and thereby $\chi =$ constant) as a function of (ξ, δ).

III. SIMPLE RADIATION MODEL

A number of important lines and bands on this diagram should be noted.

1. $T_s > T_i$, where T_i is the temperature of the first appearance of impermanent solid material at the surface, here taken as 1650 K (see Chapter 8, Section VI). This region refers to the end of the protoplanetary stage discussed immediately above.
2. $T_m < T_s < T_i$, where T_m is the freezing point (liquidus) of a dry granitic material, about 1000 K. This is the region of fractionation of the core to produce a low melting point fraction.
3. $T_s = T_m$. This is the beginning of geological time—the first appearance of potentially permanent material. The crust begins to form.
4. $T_s < T_c$, where T_c is the critical temperature of water substance. Surface liquid water appears.
5. $T \approx T_*$. The atmosphere is emptied of H_2O and CO_2, and the atmosphere and surface zone are in equilibrium with the solar input. The surface temperature no longer depends on global rearrangement.

The Net Transmissivity of the Proto-Geological Atmosphere

Taking the atmosphere as a more or less given box, the net opacity will be proportional to the amount of opaque substance present. Noting that the mass m of a constituent per unit surface area is related to the partial pressure P of the constituent at the surface by $P = mg$, then the opacity contribution of the constituent is proportional to P. Thus, for example, for the case of interest here, namely an atmosphere of water vapour, CO_2 and air

$$\frac{1}{\xi} = \left(\frac{1}{\xi_*} - 1\right)\left(\frac{P_{H_2O}}{P^*_{H_2O}} + n\frac{P_{CO_2}}{P^*_{CO_2}}\right) + 1$$

where P_{H_2O} and P_{CO_2} are the surface partial pressures of the water vapour and CO_2 gas; $P^*_{H_2O}$, $P^*_{CO_2}$ and ξ_* are reference pressures and transmissivity; n is the opacity ratio of CO_2 gas to water vapour evaluated at the respective critical points; and the unity term represents the contribution from a thin residual air atmosphere (if there is one) such that if P_{H_2O} and $P_{CO_2} = 0$ then $\xi = 1$. The choice of P^* is arbitrary—in this book I have used P^* as the critical pressure of water substance and CO_2 respectively (22.2 and 73.9 bar). Thus ξ is obtained from the two parameters ξ_* and n, both assumed to be constants of the model. (In the numerical experiments reported here I have used $n = 1$.)

It is worth noting that this is a fierce function whenever the system is dominated by water vapour, owing to the rapid variation of the saturated vapour pressure of sub-critical water substance with temperature.

There is another very important point. In reality the transmissivity is controlled only in the broadest terms by the gross amounts of the principal constituents—the roles of minor constituents, or minor dissociation products can be and often are dominant. Thus the representation of ξ given here is a gross simplification for use in a strictly phenomenological model. It proves, however, to be a very useful approach.

IV. AMOUNTS OF H_2O, CO_2

The total amount of water substance on the Earth can be estimated by totalling the amounts in the following major reservoirs. The results are expressed as an equivalent global thickness, namely volume as cold liquid water per unit area of the Earth ($5.1 \times 10^8 \text{ km}^2$).

1. Seawater, 2650 m. Obtained from the ocean volume of $1.35 \times 10^9 \text{ km}^3$.
2. Subsurface water, ≈ 300 m. Suppose the porosity e varies with depth such that $e = e_0 \exp(-z/d)$, then the equivalent thickness is $e_0 d$. Possible estimates could be $e_0 = 0.1$ and $d = 3$ km. (This reservoir includes water below both the land and oceanic solid surface. I presume an essentially dry interior below the crust.)
3. Ice (melted), ≈ 80 m. Obtained mainly from the ice caps of Antarctica and Greenland.
4. Fresh water, ≈ 40 m. Obtained from the volumes of lakes, rivers and soil water.
5. Chemically bound and biomass water, ≈ 10 m (a guess).
6. Atmosphere, 0.02 m. This is the mean precipitable water.

We obtain a grand total of 3080 m—say 3 km—the figure I shall henceforth use. Taking $g = 10 \text{ m/s}^2$, a nominal value, this gives $P_{H_2O} = 300$ bar.

The total amount of CO_2 in the Earth's proto-atmosphere is not so easy to estimate. I shall assume that this material is that currently bound in the minerals calcite $CaCO_3$ (density 2720 kg/m^3) and dolomite $CaMg(CO_3)_2$ (density 2860 kg/m^3) found principally in limestones and related rocks. Various estimates of the amount of limestones are quoted, typically as 20% of all sedimentary rocks. Let us take the following: the total "limestones" are equivalent to a 1 km thick layer over the land surface, equally as calcite and dolomite. Thus the amounts of CO_2 per unit area of the total Earth's surface are (in units 10^5 kg/m^2) 1.8 if all calcite and 2.1 if all dolomite—suggesting a nominal value of 2. If all this CO_2 is in the proto-atmosphere, the surface partial pressure $P_{CO_2} = 20$ bar.

This value is much less than the measured value on Venus (90 bar) and

IV. AMOUNTS OF H_2O, CO_2

suggests that it could be an underestimate. Nevertheless, for the Earth the proto-atmosphere will be dominated by water vapour and the model results discussed below are not grossly affected by large errors in P_{CO_2}.

A. The Amount of H_2O in the Proto-Ocean

Our model enters this stage with all the H_2O and CO_2 in a proto-atmosphere above a hot, dry surface. As the temperatures fall other phases are possible and we look for simple ways of representing them.

As an extreme possibility I assume that, once free liquid water occurs on the surface, the partial pressure of water vapour at the surface is its saturation value. Hence the equivalent depth of the proto-ocean $d = (P_{H_2O} - P_{(saturation)})/g$, provided $P_{(saturation)}(T_s) < P_{H_2O}$. Some of this liquid water will be ingested into the rock substance interior. (Also the proto-ocean is presumed to be well mixed so that the temperature difference from top to bottom is small.)

B. The Amount of CO_2 in the Proto-Ocean

In an ideal system the molar fraction of a substance in solution is proportional to the partial pressure of the gas phase (Henry's relation), so that the mass concentration n_ℓ of CO_2 in the proto-ocean is as follows:

$$n_\ell = \alpha P_{CO_2 \text{(surface)}} \qquad \alpha = \alpha(T)$$

Experimental data for $\alpha(T)$ show a mildly varying function of value $3-6 \times 10^{-4}$ bar^{-1} over 273–500 K (for details see Elder, 1981, pp. 203–205).

Thus the amount of CO_2 in the ocean per unit area is $n_\ell \rho_\ell d$. This is material taken from the atmosphere, so that

$$P_{CO_2 \text{(atmosphere)}} = P_{CO_2} - n_\ell \rho_\ell d/g$$

I have taken this into account in the numerical experiments, but the effect is very small.

C. The Amount of CO_2 in the Proto-Crust

The CO_2-based minerals are unstable at high temperatures. Some CO_2 will be dissolved in the interior but presumably the bulk will be taken up by the solid crust. The crust will have access to the CO_2 by the rapid recirculation of proto-ocean water through the crust. Thus in the model no crustal CO_2 will be allowed unless $d > 0$. I have taken the equivalent amount of $P_{CO_2 \text{(crust)}}$ to be equal to $(1 - \delta/\delta_*)P_{CO_2}$, where δ_* is a crustal thickness sufficient to take up all the original CO_2. (If the CO_2 chemically absorbed in the crust is distributed such that it decreases exponentially with depth, then δ_* is the mean depth of the CO_2 distribution. An alternative hypothesis relating $P_{CO_2 \text{(crust)}} = (1 - d/d_*)P_{CO_2}$, namely a function solely of ocean depth, is quite artificial but of interest, especially as it permits a representation of the atmosphere ocean structure without specifying the details of the crust.)

D. The H_2O, CO_2 Distribution

This latter case is particularly straightforward. Some typical results are shown in Fig. 5.5. Notice the following:

1. For T_s sufficiently high ($>T_a$, say, which in the case here is near 700 K) all the H_2O and CO_2 is in the atmosphere.
2. $T_c \leqslant T_s \leqslant T_a$. The system meets the vapour–gas line: a dotted curve is shown. Nothing actually happens in this state, rather there would be more or less discontinuous behaviour as the system passed below the critical state.
3. $T_s < T_c$. P_{H_2O} falls along the SVP line and the proto-ocean deepens. In this case the bulk of the oceanic water is deposited while the surface is still quite hot. P_{CO_2} falls, dependent on d_* which controls the amount taken from the atmosphere by the crust and ocean. For d_* very large the only effective loss is the CO_2 dissolved in the ocean. Indeed for d_* greater than a critical value ($>d_a$, say, 3 km here) the relative depletion of H_2O is greater than that of CO_2 and consequently the concentration of CO_2, namely $c = P_{CO_2}/(P_{CO_2} + P_{H_2O})$, in the atmosphere rises—so that when the system has cooled there remains a (CO_2 + air) atmosphere. For the Earth, in this model, we need $d > d_a$.

It should be noted that above the critical point of water substance ($P_c = 220$ bar, $T_c = 647$ K) no distinct liquid phase is possible. This suggests that the first appearance of the proto-ocean would be rather sudden and with an initial depth corresponding to a pressure drop of 80 bars—note, however, that near the critical point the densities of vapour and liquid are the same, about 300 kg/m³. The actual thickness of the liquid water layer d/ρ_f is indicated. The peculiar shape arises, including the dip immediately below T_c, as a consequence of the thermodynamic nature of water substance. What is most striking is that the water layer thickness varies so little with surface temperature and that it is initially already about 2.5 km.

The discussion of the circumstances of the initial geological atmosphere and crust has concentrated attention on its thermodynamic structure. Now let us consider how long this stage takes.

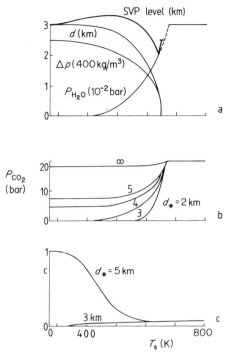

Fig. 5.5. Proto-atmosphere–ocean-crust properties as a function of temperature at the base of the atmosphere T_s (K), for illustrative model with $P_{CO_2(crust)}$ a function solely of ocean depth d. (a) H_2O properties. (b) CO_2 pressure. (c) CO_2 concentration in the atmosphere.

V. THE PROTO-CRUST

When the surface temperature has fallen below that of the liquidus, for the material in the surface zone simple "freezing" can occur from the surface downwards. The material of this solid layer will be less dense than the liquid from which it forms and will be at temperatures below the liquidus—it is thermodynamically stable. Other processes can still rearrange it, e.g. erosion from the action of the hydrosphere and internal chemical or mechanical processes. Nevertheless, there is the possibility, even if a very slight one, that some of this new layer could be preserved. Geological time has commenced.

The essence of the thermodynamics of the layer is the release of latent heat from fluid at the interface at the base of the layer as growth proceeds by freezing on more material. If the process is not too rapid, so that the temperature distribution in the layer is in a quasi-steady state, then the heat flux through the layer $f \approx \rho c \kappa \Delta\theta / y$, where $\Delta\theta = T_m - T_s$, the temperature difference across the layer of thickness $y = y(t)$. If this heat comes solely from the release of latent heat, $f = \rho L \, dy/dt$, where L is the specific latent heat. Hence we have

$$dy^2/dt = (2c\kappa/L)\,\Delta\theta$$

In the simple case for which $\Delta\theta = $ constant

$$y^2 = (2c\kappa/L)\,\Delta\theta\, t$$

and the layer thickness increases as $t^{1/2}$. For example, with $L = 4 \times 10^5$ J/kg, $c = 10^3$ J/(kg K), $\kappa = 10^{-6}$ m², $T_s = 300$ K, $T_m = 1000$ K, then $y = 10.5(t/\text{year})^{1/2}$ m so that $y \approx 0.1, 1, 10$ km in $10^2, 10^4, 10^6$ year. (See Carslaw and Jaeger, 1959, chapter 11.)

Here, however, $\Delta\theta = T_m - T_s(t)$ is an increasing function of time, with a time scale of order 10^5 year.

Whereas the time scale of the pre-crustal stage is determined essentially by the energetics of the fractionation, the temporal behaviour during this stage is determined by the freezing process.

In the numerical model there is the minor problem of the relation of the surface slag and the frozen layer. Quite arbitrarily I have set $y = 0$ at $t = 0$, namely when first $T_s = T_m$, and taken the thickness of the crustal barrier $\delta = \delta_{(\text{slag})}$ while $y < \delta_{(\text{slag})}$ and $\delta = y$ when $y \geqslant \delta_{(\text{slag})}$. Since $\delta_{(\text{slag})}$ is taken here as no more than 100 m, this could lead to a possible error in time of order 10^2 year. In view of the time scale of the whole stage this is negligible.

Also we must not forget that there is an upper limit on y arising from the total amount of potential granitic material available. At most this is about equivalent to a global layer of 10 km thickness.

VI. MODEL OF THE EARLY ATMOSPHERE–OCEAN–CRUST FOR THE EARTH

A number of powerful processes have been identified: the fractionation prelude; a radiation model with its associated transmissivity relation; a model for the distribution of H_2O and CO_2 in the atmosphere–ocean–crust; and a freezing model for the growth of the proto-crust. When combined together they provide a model which gives, as it were, a thermodynamic glimpse of the circumstances at the time of the onset of the geological stage.

There is a lot of room for choice of the parameters of the model. I have given some indication of how I have chosen them and how different values will alter, slow down or speed up the process. All I present here is a scenario

VI. MODEL OF EARLY ATMOSPHERE–OCEAN–CRUST

which is a reasonable choice of parameters. Let us look at the model results shown in Fig. 5.6.

1. $0 \leqslant t \leqslant 1.2$ Ma. This is the fractionation stage. The surface temperature falls from 1650 to 885 K. All the volatiles are in the hot atmosphere.
2. $t > 1.2$ Ma. This is the initial part of the geological stage. A crust forms; the atmosphere drops its H_2O to form an ocean; CO_2 is absorbed into the ocean and crust.
3. $1.2 < t < 1.4$ Ma. the atmosphere rapidly drops its H_2O once $T_s < T_c$ and there is a partial increase in transmissivity. An ocean forms. Keep in mind that d is the equivalent depth.
4. $1.2 < t < 1.5$ Ma. The atmosphere loses its CO_2. Note that this interval is determined by d_* (here 5 km).
5. $1.2 < t < 1.8$ Ma. The transmissivity of the atmosphere increases and takes an extended period to completely clear owing to the residual CO_2 (a function of ξ_*, 10^{-4} here).
6. $t > 1.8$ Ma. The proto-geological stage is over. The atmosphere has only air and minor H_2O and CO_2; the ocean is full; all the granitic material is in the proto-crust; the surface temperature is steady and no longer influenced by internal processes.
7. $1.2 < t \lesssim 2$ Ma. The proto-crust grows until limited by the availability of granitic material.

(Measuring time t' from the moment $T_s = T_m$: $t' \approx 450$ year, proto-crust begins; $t' \approx 2.5 \times 10^4$ year, liquid water begins to form on the surface; $t' = 4 \times 10^5$ year, $T_s \approx 320$ K, ocean fully formed, crustal thickness ≈ 6 km.)

This is a sequence of phase changes: fractionation of low melting point rock substance; crustal slag separation; H_2O vapour to liquid H_2O; CO_2 gas to dissolved CO_2; and freezing of granitic material, which arises as the body cools. This leads to a sequence of relatively sharp changes in the behaviour of the system as it crosses major phase boundaries. The time scales of the various stages are controlled by the role of the atmosphere, through the parameter ξ and by the internal processes of fractionation and freezing through the parameter δ.

The planet at this stage has a liquid rock substance interior contained within a thin, fragile skin surrounded by a thin atmosphere. At long last it is a truly distinct entity ready to commence its geological life.

The Role of Hydrothermal Activity

There is an important feature as yet left aside. Once free liquid water is available to the crustal system, vigorous hydrothermal activity is possible in

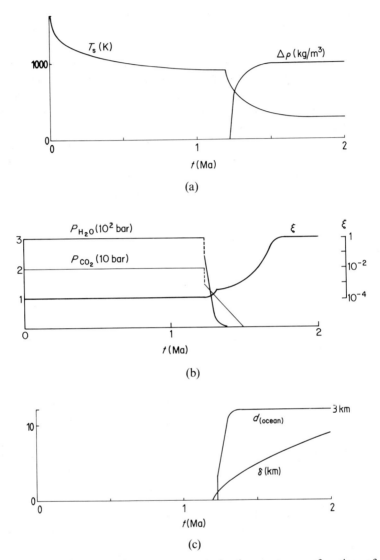

Fig. 5.6. Earth: atmosphere–ocean–crust development as a function of time t (Ma). (a) Temperature T_s (K) at base of atmosphere. The density contrast, $\Delta\rho$ (kg/m^3), between liquid and vapour water substance is also shown. (b) Atmosphere partial pressures P_{H_2O} and P_{CO_2} (bar) and the transparency ξ. (c) Ocean depth d (km) and crustal thickness δ (km). Model parameters: $T_i = 1650$ K; $T_m = 1000$ K; $T_0 = 280$ K; $P_{H_2O} = 300$ bar; $P_{CO_2} = 20$ bar; $\xi = 10^{-4}$; $\delta_{(slag)} = 100$ m; $T_{(surface\ liquid\ onset)} = 647$ K.

VI. MODEL OF EARLY ATMOSPHERE–OCEAN–CRUST

which water is rapidly circulated through the crust. This will speed up the process of crustal formation by a factor $N^{1/2}$, where N is the average Nusselt number of the global hydrothermal systems—taking the effect as enhancing κ to $N\kappa$. Vigorous modern hydrothermal systems would suggest $N \sim 10^2$ (see for example Elder, 1981) so that the proto-crustal stage, once liquid water is available, can proceed about 10 times faster than the process described so far.

In order to give a quantitative description we need to be able to evaluate the heat transfer capability of the global hydrothermal systems—to be able to state the depth of the crust affected by the hydrothermal activity and to estimate the effective thermal conductivity, where N is the Nusselt number of the aqueous convective system in the permeable crust. This would, for example, involve some knowledge of the variation of permeability with depth. Even guesswork is of little help, so I have taken a wantonly empirical approach, after a lot of fiddling, namely:

$$N_0 = N_* \phi^3 \qquad \phi = (T_{(\text{critical})} - T_s)/T_{(\text{critical})}$$
$$y \leqslant h_0: \quad N = 1 + N_0 y/h_0$$
$$y > h_0: \quad N = 1 + N_0(h_0/y)^4$$

provided $T_s < T_{(\text{critical})}$, else $N = 1$, where y is the crustal thickness and N is determined by the two parameters h_0, a measure of the depth of circulation of liquid water, and a reference Nusselt number N_*. In the model results shown in Fig. 5.7 I have taken $h_0 = 3.5$ km, about the depth at which the BPD (boiling point for depth relation of a volume of liquid water everywhere at the boiling point) reaches T_c; and $N_* = 300$, representative of modern active hydrothermal systems, and a value such that when the crust is fully developed N is not too much greater than unity.

In a cartoon like this the precise details should not be taken seriously. Nevertheless we see a dramatic alteration in the behaviour of the system.

1. The most striking effect is the pronounced hydrothermal pulse, as shown by N, in the interval 50–100 ka.
2. There is, however, little effect on the development of P_{H_2O} since the atmosphere is depleted before the hydrothermal activity gets going.
3. The effect on the crustal development is pronounced. From about 60 ka its growth is accelerated as heat is rapidly extracted from the interior by the hydrothermal activity.
4. There is a correspondingly more rapid fall in surface temperature as a consequence of the larger δ. The equilibrium surface temperature is reached after 80 ka.

It would be of great interest to linger here during this very busy part of

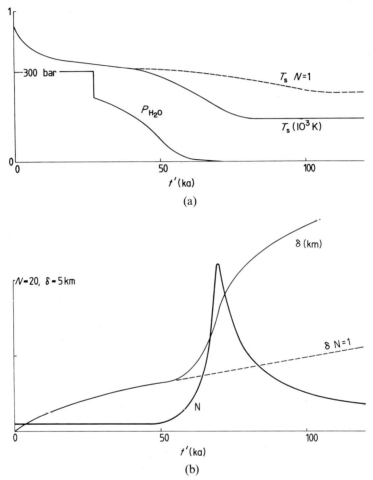

Fig. 5.7. Earth: atmosphere–ocean–crust development, allowing for hydrothermal activity, as a function of time t' (ka). (a) Temperature T_s (K) at base of atmosphere, the corresponding curve for $N = 1$ shown dashed for comparison. Atmosphere partial pressure P_{H_2O} (bar). (b) Crustal thickness δ (km), the corresponding curve for $N = 1$ shown dashed for comparison. Nusselt number N.

proto-geology but we also need to keep an eye on what is happening elsewhere in the solar system.

VII. PLANETARY PERSPECTIVE

For the Earth, as we have seen, the role of the proto-atmosphere is very strong in the pre-geological stage of a solid planet's development. Clearly, however, many of our ideas are guesswork. There are few clues of value.

VII. PLANETARY PERSPECTIVE

Shown in Table 5.1 are the properties of the common volatiles and in Fig. 5.8 the range over which some of these are liquid, on a plot of mean temperature as a function of solar distance, namely $T/T_{\text{(earth)}} = (r_{\text{(earth)}}/r)^{1/2}$ with $T_{\text{(earth)}} \approx 280\,\text{K}$, together with the band of temperature ranging from $2^{1/2}T$, the sub-solar temperature, to $2^{-1/2}T$.

It is immediately clear that both Earth, the H_2O planet, and Titan, the CH_4 moon of Saturn, are massive enough to retain a substantial atmosphere, but that the dominant atmospheric constituent is close to its melting point at the surface temperature. In other words, methane plays the same role on

Table 5.1. Properties of common volatile substances

	MP (at 0 bar) (K)	BP (at 1 bar) (K)	T_c (K)	P_c (bar)	Comment
H_2O (18)	273.16	373.15	647.3	221.2	Liquid range 100 K $\rho_c = 315\,\text{kg/m}^3$
NH_3 (17)	195.5	239.8	405.5	112.8	Liquid range 44.3 K $\gamma = 1.31$ $\rho_c = 810\,\text{kg/m}^3$
CO_2 (44)	216.6	(194.7)	304.2	73.9	Sublimes $\rho_c = 240\,\text{kg/m}^3$
CH_4 (16)	90.7	109.2	190.7	47.7	Small liquid range of 18.5 K $\gamma = 1.31$ $\rho_c = 630\,\text{kg/m}^3$
O_2 (32)	54.8	90.2	154.8	50.8	$\rho_c = 410\,\text{kg/m}^3$
N_2 (28)	63.3	77.4	126.2	33.9	$\rho_c = 400\,\text{kg/m}^3$

Titan as does water on Earth—with the lower atmosphere energetics being dominated by the transport of latent heat between the liquid and vapour phases. This situation does not occur on any other planetary body—either the body is of too small a mass to retain a substantial atmosphere, as is the case for Mars, or an appropriate volatile does not exist or is accessible only as a frozen solid, as in the case of the moons of Uranus and Neptune.

To provide, then, some sort of answer to the outstanding question here, namely how it is that Venus has a thick, hot atmosphere and the Earth not, we see that it is all to do with the availability of water substance. On the Earth, once the surface temperature falls below T_c, condensation can occur, liquid H_2O can transport CO_2 into the rock substance interior to form calcite/dolomite, and the atmosphere is depleted in both H_2O and CO_2. This cannot happen for Venus—there is no H_2O and the CO_2 must remain in the atmospheric trap to provide a dense opaque atmosphere.

If there had been substantial water substance on Venus a similar development to that described for the Earth would have been possible. Plainly there

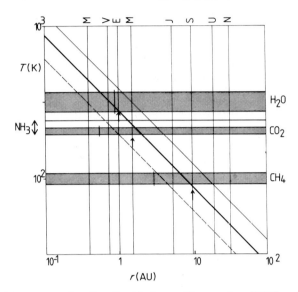

Fig. 5.8. Occurrence of surface liquid phases. Temperature T (K) and temperature range on the surface of the planets; and the melting–boiling range for various substances.

has been no such development. We are forced to the conclusion that either Venus never had any water substance or that if it did then it has all been lost.

On the question of the availability of water substance, we may consider the possibility that owing to differing thermodynamic conditions in the nebula the concentration of chemical species also differed. This is undoubtedly true but it is most unlikely that the concentrations varied so strongly that there was no H_2O for proto-Venus. Indeed the temperature of the solar front remains nearly constant as it sweeps inward so that the residual material left behind should have had a nearly uniform composition. It is very likely that water substance was available.

Water substance must then have been lost. Presumably this would be by photo-dissociation and escape of high-level gaseous H_2O in the early envelope of the proto-planet. This being so the process must have been extreme for proto-Venus, moderate for Earth and Mars, and negligible for the Jovian bodies and their satellites.

Thus the major constituents of a planetary atmosphere are those minor residual components which remain after the major H-degassing stage and the subsequent loss from the proto-atmosphere either by dissociation and escape or by condensation out of the atmosphere.

VIII. MODEL FOR VENUS

The (ξ, δ) diagram for Venus is shown in Fig. 5.9. In addition to the lines of constant T_s, there are lines of constant f_i, the flux from the interior. Since we are now presuming that Venus has not condensed any matter from the proto-atmosphere, then $\xi = $ constant through its development. An appropriate value is $\xi \approx 10^{-5}$. This produces a proto-geological endpoint, when all the granitic material about 6 km thick is locked up in the proto-crust, at $T_s \approx 700$ K, with $f_i \approx 0.13$ W/m^2.

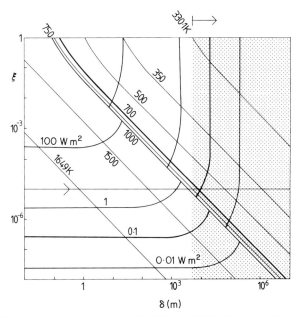

Fig. 5.9. Venus: transparency, crustal thickness (ξ, δ) diagram. Lines of constant surface temperature and heat flux from the interior (W/m^2). Suggested development along $\xi = 10^{-5}$. In the shaded region the granitic source is exhausted.

It is worth emphasizing in passing, as highlighted by Fig. 5.9 in the interval 1500–500 K, that for the interior heat flux there is a transition from interior heat flow dominated by ξ to the interior heat flow being independent of ξ— a transition from dominance of the proto-atmosphere to dominance of the proto-crust. Venus remains in this transition zone.

The model gives the following results: fractionation stage T_m reached in 6.2 Ma, permanent crust forming after 8.1 Ma and full granitic proto-crust of 5 km thickness in 8.9 Ma with a surface temperature of 735 K.

IX. MODEL FOR MARS

A. The Amounts of H_2O and CO_2 in the Proto-Atmosphere

The direct evidence for the existence of H_2O and CO_2 on Mars is from the composition of the atmosphere and the existence of the polar caps. The atmosphere is largely composed of CO_2 vapour at a pressure of about 10^{-2} bar—not too different from the SVP of vapour in equilibrium with solid CO_2 at the mean surface temperature of 225 K. A minor atmospheric constituent is H_2O vapour at a pressure in equilibrium with ice at 200 K. The south polar cap is believed to be largely dry ice; the north polar cap is believed to be largely water ice.

Presumably all the primeval CO_2 has been retained. Some of the primeval H_2O will have been lost after photo-dissociation in the atmosphere. We can envisage a near-surface permafrost zone extending possibly to depths of order 1 km in which the rock voids are filled with frozen H_2O and CO_2. Outside the polar regions where the temperatures are higher there is likely to be a zone of depth of perhaps of order 10–100 m which has become depleted of both H_2O and CO_2, but especially of H_2O under the strong daily and annual thermal cycling.

There is no evidence to indicate the actual amounts of these two constituents. Mars is, however, a much less massive body than the Earth, so that the total amounts, even allowing for the somewhat lower surface temperature of Mars, should be much less. Model values of initial proto-atmosphere amounts are taken as $P_{H_2O} = 10$ bar and $P_{CO_2} = 10$ bar. (These are just nominal values. The argument above suggests that in reality $P_{CO_2} > P_{H_2O}$.)

Thus most of the H_2O and CO_2 today is locked up in the crust. The atmospheric pressure of these constituents will be determined by a meteorological equilibrium between inputs from sublimation and outputs as surface deposits at pressure levels below the SVPs of the mean surface temperature.

B. Model Sequence

Thus we envisage a sequence of events similar to that on Earth with the additional phase changes: (1) liquid H_2O freezing; and (2) solid CO_2 forming.

In the case of the Earth the CO_2 has been taken up in calcite/dolomite minerals under the action of hot liquid H_2O. Such a process is inhibited under the conditions on Mars—but there is the sink for CO_2 as dry ice deposits. So we have the interesting analogue of limestones on Earth and dry ice sediments on Mars.

IX. MODEL FOR MARS

C. Allowance for Non-Uniform Surface Temperatures

The solar flux per unit area of a planet varies from equator to pole because of the range of angle of incidence of the radiation on the surface. This effect is enhanced for planets with substantial inclination of the axis of rotation (range for Mars 14°–35°). A thick, vigorously convecting atmosphere smooths out the effect on the surface temperature. This is not so when the atmosphere is thin. In the model here this is not usually of much consequence. But in the case of Mars there is the fortuitous circumstance that the mean surface temperature, $T_s \approx 225$ K, is rather close to the boiling point, BP = 216.6 K, of the major atmospheric constituent, CO_2. It would be unrealistic not to take this into account.

Since this situation arises only in the very last part of the proto-geological stage as the surface temperature approaches its value in equilibrium solely with the solar input, a very simple representation is adequate. Thus, as sketched in Fig. 5.10, the surface temperature ranges from, say, $T_a = r_a T_s$ to

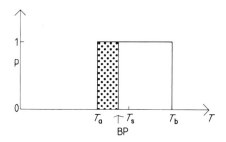

Fig. 5.10. Surface temperature probability function.

$T_b = r_b T_s$, with BP $= sT_s$; and assuming a "top-hat" probability distribution for temperature (in reality a more or less truncated cosinusoidal-like function of latitude and inclination) then the fraction f_T of CO_2 in the liquid/solid state is as follows: MP $< T_a, f_T = 0,$ all the CO_2 is above the BP as a vapour; MP $> T_b, f_T = 1$, all the CO_2 is below the BP as a liquid/solid; and $T_a <$ MP $< T_b, f_T = (s - r_a)/(r_b - r_a)$.

In the model I have rather arbitrarily taken $r_a = 0.8$ and $r_b = 1.2$, half the range that would be suggested from the ratio $T_{\text{(sub-solar)}}/T_* = 2^{1/2}$. Furthermore, one needs an estimate of the effective fractional area f_a of the planet's surface which contains permanent solid CO_2 deposits. A value of $f_a = 10^{-2}$ is suggested by the atmospheric CO_2 pressure today.

D. The CO_2 Sediments

Most of the CO_2 on Mars is presumably trapped in "dry ice sediments". As a reasonable guess, these sediments are finely and irregularly layered and weakly cemented rocks in which loess and dry ice layers are intercollated. The loess arises from the windy dusty atmosphere and the dry ice by reverse sublimation, also from the atmosphere.

The rate at which this material was originally deposited can be represented as follows. Suppose the rate of accumulation of dry ice in the sediments, meaning the amount as equivalent to a pressure $P_{(seds)}$, is proportional to the pressure of CO_2 in the proto-atmosphere P with a probability of occurrence f (see above). Hence

$$dP_{(seds)}/dt = fP/\tau$$
$$P = P_0 - P_{(seds)}$$
$$d_{CO_2} = P_{(seds)}/\rho g, \text{ the global mean thickness of dry ice}$$

provided $P > f_a f_T$. SVP where τ is the time scale of the process. Until direct inspection of the deposits is possible an estimate of τ is little more than guesswork—on Earth, Recent loess deposits of thickness of order 10^2 m have been produced in intense conditions in times of order 10^4 year, the value assumed here for τ. In any event, especially for small values of transmissivity, the precise value of τ has little overall effect on the interval for this stage.

E. Properties of CO_2

The model uses the following values (Weast, 1983): $\rho_{(dry\ ice)} = 1250\ kg/m^3$, as compacted "cake"; and $\rho_{(liquid)} = 1180\ kg/m^3$, at MP = 216.6 K. (1 m of dry ice, on Mars with $g = 3.72\ m/s^2$, gives a pressure change of 0.0465 bar—in other words 21.5 m/bar.)

For temperatures below the melting point at 216.6 K the saturated vapour pressure SVP of CO_2 vapour in equilibrium with solid CO_2 is given by the empirical relation:

$$\log_{10}(P/\text{bar}) = 7.03 - 1370/(T/K)$$

This gives $P \approx 1$ bar at the boiling point 194.7 K.

The relation can be used to obtain the SVP for temperatures somewhat about the melting point—e.g. at 225 K, the nominal mean surface temperature of Mars, we have SVP = 9.7 bar.

F. Mars Model Results

Typical model data are shown in Figs 5.11 and 5.12. This is the most complex situation because of the many phase changes: fractionation, formation of

IX. MODEL FOR MARS

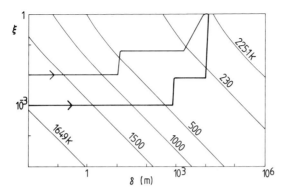

Fig. 5.11. Mars: transparency, crustal thickness (ξ, δ) diagram. Lines of constant surface temperature and two possible development paths with initial $\xi = 10^{-3}, 10^{-2}$.

slag and permanent crust, together with the condensation of two volatile components. Because of the wide range of time scales the data of Fig. 5.12 is plotted against t', the time since the nominal end of fractionation. I shall describe the results for the case $\xi_* = 10^{-3}$.

1. Fractionation has a time scale 170 ka, the surface temperature falls from 1650 to 1000 K in 62 ka, and the stage ends after 470 ka with $T_s = 650$ K. Subsequent events occur soon after.
2. $t' = 0.25$ ka, permanent crust forms.
3. 7.5 ka $\lesssim t' \lesssim$ 35 ka. Liquid H_2O forms on the surface, accumulating

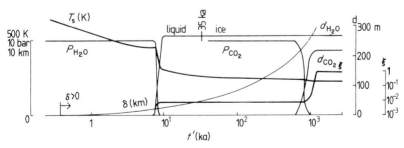

Fig. 5.12. Mars: atmosphere–ocean–crust development as a function of time t (ka). Temperature, T_s (K); atmospheric partial pressures, P_{H_2O} and P_{CO_2} (bar); transparency, ξ; equivalent depth of water substance (liquid to ice change at $t = 35$ ka) and dry ice (m); crustal thickness, δ (km). Model parameters: $T_i = 1650$ K; $T_m = 1000$ K; $T_0 = 280$ K; $P_{H_2O} = 10$ bar; $P_{CO_2} = 10$ bar; $\xi = 10^{-3}$; $\delta_{(slag)} = 100$ m; $T_{(surface\ liquid\ onset)} = 647$ K.

to an equivalent depth of 270 m, and then becomes frozen. In this model there is always an interval of liquid surface H_2O, but it is always a brief one, here barely 30 ka. The persistent suggestion that the morphology of the Martian surface is a relic of fluvial processes would therefore seem most unlikely—the vigour of aeolian transport, together with internal rearrangement, is a much better candidate. The bulk of the H_2O would have been retained in the regolith and near-surface rocks as pure H_2O and then become permafrost. The atmosphere loses its H_2O but remains opaque in this interval and the subsequent fall of surface temperature is slow.

4. 750 ka $\gtrsim t' \gtrsim$ 1000 ka. Dry ice is deposited and dry ice sediments accumulate. The atmosphere clears, is maintained by sublimation, and remains in a dynamical equilibrium with the near-surface deposits of ice and dry ice. The permanent crust reaches a thickness of about 10 km after 900 ka.

X. MODEL FOR TITAN

The pre-geological development of Titan is similar to that of Earth in that it is dominated by a single volatile substance, CH_4, rather than H_2O. Model data are shown in Figs 5.13 and 5.14. Time is measured from the moment the surface first freezes.

(Methane properties in Table 5.1: also $\rho_\ell = 400 \text{ kg/m}^3$ at 1 bar, 273 K; $\rho_v = 0.72 \text{ kg/m}^3$ at 1 bar, 273 K (Weast, 1983). Note that on Titan there is a pressure change of 1 bar down a liquid methane column of height 180 m. Ice

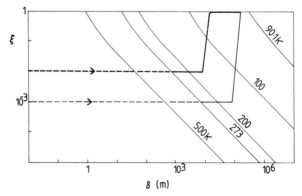

Fig. 5.13. Titan: transparency, crustal thickness (ξ, δ) diagram. Lines of constant surface temperature and two possible development paths with initial $\xi = 10^{-3}, 10^{-2}$.

X. MODEL FOR TITAN

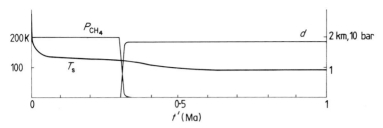

Fig. 5.14. Titan: atmosphere–ocean–crust development as a function of time (Ma). Temperature, T_s (K); atmospheric partial pressure, P_{CH_4} (bar); methane ocean depth, d (km). Model parameters: $T_i = 373$ K; $T_m = 273$ K; $T_0 = 90$ K; $P_{CH_4} = 10$ bar; $\xi = 10^{-2}$; $\delta_{(slag)} = 10$ m; $T_{(surface\ liquid\ onset)} = 124$ K.

properties at 90 K: $K = 6$ W/(m K), $\rho = 935$ kg/m^3, $L = 200$ kJ/kg, $\kappa = 9.4 \times 10^{-6}$ m^2/s (Hobbs, 1974). $M = 1.36 \times 10^{23}$ kg, $a = 2560$ km, $g = 1.4$ m/s^2.)

1. $t' < 280$ ka. The surface temperature falls, rapidly at first, and the (ice) crust grows to a thickness of 9.5 km. The atmosphere is unchanged.
2. 280 ka $< t' <$ 380 ka. Liquid methane condenses from the atmosphere to form an ocean of equivalent depth 1785 m. (Methane ice will form caps near the polar regions and methane bergs will be found even in equatorial regions—the temperature range MP to BP is only 18.5 K.) The atmosphere is then fully depleted. Only the residual N_2 will remain (temperatures are not low enough to condense the N_2), together with CH_4 in dynamical equilibrium with the surface material.

This is a truly science fiction type of planetary body—exceedingly chilly yet with a very active "methosphere" and vigorous methogeomorphology.

The phase changes which dominate the proto-geological stage occur quickly— the amounts of substance are small; the temperature ranges between one phase and the rest are small. The subsequent development is slow because the amounts of substance are large, as is the temperature range between phases. It has taken a time of order 1 Ma to set the geological stage; it requires a time of order 10^3–10^4 Ma for the geological drama itself.

CHAPTER 6
Chemical Development

I. INTRODUCTION

The geological clock is running. Already the prelude of global fractionation, the segregation of the low melting point materials into the granitic crust, is complete. The liquid rock substance interior continues to act as a pool or reservoir from which a succession of distinct solid phases are formed to produce a growing solid mantle built from a sequence of distinct chemical and mineralogical phases.

A key feature of the model of global fractionation described here is that it is calibrated against data from upper mantle rocks. This is the major geological feature of the model: the keystone of this and the following two chapters.

II. FRACTIONATION EVIDENCE

The terrestrial planets are not chemically homogeneous. There is direct evidence from the Earth. The crust is very heterogeneous: mantle samples obtained as xenolithic fragments in erupted material, kimberlites etc. are different chemically from crustal material; the seismic identification of the elastic properties and density variation suggest substantial radial heterogeneity, especially across the core–mantle boundary.

This chemical variation can be thought of as arising from a fractionation process in which new mantle material is obtained as an extract from the pool of core liquid. In this view the new mantle material will differ chemically from that of the pool since some chemical species will have a preference for the solid extract.

The fractionation process will be represented quantitatively here solely through the "fractionation ratio", the molecular ratio of a species in the solid extract with that in the pool liquid. These ratios will be assumed to be constants independent of the thermodynamic state—a strong assumption. The species of interest are the major constituents of rock substance and they will be represented as oxides. Whether or not this representation as oxides is physically realistic is arguable—it raises many questions about the atomic arrangement of liquid rock substance and about the details of the freezing process and the rearrangements of the constituents into distinct minerals.

The species will be given the following symbols for a single (hypothetical) molecule: $f^* = Fe_2O_3$, $f = FeO$, $m = MgO$, $c = CaO$, $n = Na_2O$, $k = K_2O$, $a = Al_2O_3$ and $s = SiO_2$. Thus a single molecule of anorthite has the symbol cas_2. Note that, in analyses quoted here, owing to uncertainty of the oxidation state of Fe all Fe measured as Fe_2O_3 has been "converted" to FeO. The symbols f–s will also be used to represent molecular ratios—this will be clear from the context.

A. The Primary Material

A fundamental assumption of this work is that the (mean) chemical composition of the total solid residues of the proto-solar system (now contained in the terrestrial planets, their moons, asteroids, meteoroids etc.) is identical to that of the Sun except for the volatiles (principally hydrogen and helium). Although I shall make the additional equally strong assumption that the proto-solar system was grossly chemically homogeneous—so that the mean chemical composition of the rock substance of the terrestrial planets is the same—we realize that powerful processes of chemical differentiation were active during the evolution of the proto-solar system. The hydrogen-flush mechanism is an example of such a process. Also, as is well appreciated, Mercury has an "anomalous" mean density for a planet of its size. Nevertheless, let us see for the moment how far we can go on the assumption that all the terrestrial planets are made of the same material.

The constitution of the primary material has been shown in Table 1.1. The outstanding feature revealed by this data is the simple nature of the primary material composed largely of m, f and s. (Less than 5% is as c, n, k and a, the constituents most commonly occurring in feldspars.)

Note that the oxide molecular concentrations, stated as parts per 1000, used in the numerical experiments reported in this book are given in Table

II. FRACTIONATION EVIDENCE

1.1. The only departure from these values is in the choice of various value for f. I do *not* renormalize the concentrations to give a total of 1000. To obtain normalized values divide all the concentrations, f also, by $(1 + 0.001(f - 316))$.

B. Upper Mantle Fractionation Ratios

Let us now look at the composition of suites of crustal and upper mantle rocks in order to see that fractionation occurs and to quantify it. Various compositions are given in Table 6.1—the data columns being in order of decreasing s.

Table 6.1. Oxide molecular ratios for various Earth rocks

	UCC	GC	Andesite	Basalt	glh	U
			oxides: molecular ratios			
f	42	58	64	84	51	61
m	38	57	48	106	487	527
c	47	77	81	133	33	20
n	35	34	39	25	4	2
k	24	15	11	3	2	0.1
a	96	100	110	100	16	13
s	718	659	647	549	407	377

See footnote to Table 6.2 for sources of data.

Comparing the composition of the primary material with that of the upper mantle we obtain estimates of the molecular fractionation ratios shown in Table 6.2. Note that these ratios are normalized to the fractionation ratio of silica, $k(s) \equiv 1$. The key feature of this data, keeping in mind the dominance of f, m and s, is that $k(f)$ is small and $k(m)$ somewhat greater than unity.

It is of immediate interest to note also that the ratio f/m is roughly unity in the crust but is an order of magnitude smaller in the upper mantle—owing to the dominance of m. The f/m ratio is also roughly unity in the Sun. This strongly suggests a fractionation ratio for f of about 0.1.

The consistency of the data is surprisingly good considering that it is obtained somewhat arbitrarily. The spread of values is 20% or less. The only awkward value is that for k with two very different estimates (0.1, 1.5)—a value similar to that of n might be expected—but fortunately this is for material of low abundance which has negligible effect on the behaviour of the global system (see below where the fractionation ratio for k is taken as 1, no fractionation).

Table 6.2. Fractionation factors for the Earth's upper mantle

	Ultramafics			Garnet lherzotites		Mean		Model search range	Preferred value
	α	β	(a)	β	(b)				
f	316	61.2	0.18	51.2	0.14	0.16 ± 0.02	(12%)	0.1–0.2	0.15
m	296	527.6	1.63	487.5	1.40	1.5 ± 0.1	(7%)	1.2–1.7	1.5
c	21	20.4	0.90	33	1.35	1.1 ± 0.2	(18%)	0.7–1.5	1.0
n	10	2.1	0.2	3.5	0.3	0.25 ± 0.05	(20%)	0.15–0.35	0.3
k	1	0.1	0.1	1.7	1.5	?		0.1–2(?)	1.0
a	12	12.7	0.98	15.5	1.11	1.05 ± 0.05	(5%)	0.95–1.15	1.0
s	345	376.8	≡1.00	407.5	≡1.00	≡1.0		≡1	≡1.0

The data for Tables 6.1 and 6.2 have been taken from the following works:

UCC, upper continental crust. Wederpohl (1975, p. 65).

GC, global average crust. Assumes GC = $(2a + b + c)/4$, where a = UCC, b = lower crust of andesite (above) and c = oceanic crust of basalt (above), roughly for 70% area of oceanic crust of 5 km basalt, 30% area of continental crust of 20 km UCC and 10 km andesitic (above) lower crust.

Andesite (p. 11); basalt (pp. 11, 33 #2); glh, garnet–herzolite (p. 625 #3). Carmichael et al. (1974).

Ultramafics, average of 168 rocks. White (1967).

(a) and (b) are ratios β/α, scaled to give s factor of unity.

All Fe given as Fe_2O_3 in the above analyses has been converted to FeO.

III. THE MINERAL ASSEMBLAGE

The working material of the model to be developed here is a collection of atoms represented by the molecular proportions of their oxides. In the liquid state this is an adequate representation. In the solid state, however, the material forms rock characterized by a particular mineral assemblage. In principle, to determine the assemblage from a given set of oxides requires, as well as a knowledge of the phase space of the substance, a statement of the temporal thermodynamic path through the phase space—the formidable task of experimental petrology.

Various schemes have been produced to short-circuit the enormous complexity and relative ignorance of multicomponent systems by generating a set of "normative" minerals from a given set of oxides. The best known of these schemes, the CIPW scheme (see, e.g., Carmichael *et al.*, 1974, and the relevant references therein), exploits the observation that when a magma is frozen the minerals tend to appear in a fairly definite order. Thus if a set ordering of the minerals in their mutual competition for the oxides is given it is straightforward to calculate the "norm". The CIPW scheme is actually quite complicated in the manner in which it copes with non-standard material and various special circumstances, but these matters are of no concern here since the system of interest is largely an (f, m, s) system with only two dominant normative minerals, olivine and hypersthene.

Normative Mineral Model (CIPW Scheme)

Given the molecular proportions of the oxides (f^*, f, m, c, n, k, a and s), the normative minerals—indicated in Table 6.3—are generated in three main steps, which in outline are as follows:

1. Form in order: or, ks, ab, an, cm, ac, ns, mt, hm, di, hy, wo. Some of these will be zero. The amount of silica remaining is taken as qz. If qz $\geqslant$ 0 there is nothing more to do. Otherwise proceed to the second step.
2. Reconstitute some of the normative minerals, formed in the first step, successively until the excess silica is no longer negative, that is qz = 0. The order of reconstitution is:

$$\begin{aligned} \text{hy:} &\quad \rightarrow \text{hy, ol} \\ \text{ab:} &\quad \rightarrow \text{ab, ne} \\ \text{or:} &\quad \rightarrow \text{or, le} \\ \text{di:} &\quad \rightarrow \text{di, ol} \\ \text{le:} &\quad \rightarrow \text{le, kp} \end{aligned}$$

Table 6.3. CIPW normative minerals (selection)

Abbreviation	Name	Formula	Molecular weight
qz	Quartz	s	60
cm	Corundum	a	102
or	Orthoclase	$ka\,s_6$	556
ab	Albite	$na\,s_6$	524
an	Anorthite	$ca\,s_2$	278
le	Leucite	$ka\,s_4$	436
ne	Nepheline	$na\,s_2$	284
kp	Kaliophite	$ka\,s_2$	317
ac	Acmite	$nf^*\,s_4$	462
ns	Sodium metasilicate	ns	122
ks	Potassium metasilicate	ks	156
di	Diopside	$c(m,f)s_2$	217–249
wo	Wollastonite	cs	116
hy	Hypersthene	$(m,f)s$	101–132
ol	Olivine	$(m,f)_2 s$	140–204
cs	Calcium orthosilicate	$c_2 s$	172
mt	Magnetite	ff^*	232
hm	Hematite	f^*	160

3. If excess silica is still negative—as can happen with dominant $m' = (m,f)$—reconstitute the m',s minerals hy and ol in part or wholly to their constituent oxides:

$$\text{hy:} \rightarrow \text{hy}, m'$$
$$\text{ol:} \rightarrow \text{ol}, m'$$

As used in this work, the following minor modifications have been made:

1. Not knowing the valency state of Fe all Fe is "converted" to $f =$ FeO.
2. Fe^{+3} minerals: acmite, magnetite and hematite are not formed.
3. As usual the other normative minerals do not distinguish between m and f, these oxides being regarded as fully interchangeable, so that $m' = (m,f) = m + f$ is used—e.g. ol $= m'_2 s$.
4. Any excess m' remaining is given the symbol sp $= (m,f)$ in the diagrams.

The results of this procedure is shown in Table 6.4 for the rock substances of interest here. Note that if the solar material were formed into an igneous rock it would be composed largely of olivine—a delightful observation. Included are our reference rock substances and, anticipating the work of this chapter, the model rock substance at the top of the Earth's mantle.

IV. POOL MODEL 113

Table 6.4. Normative minerals of various rock substances: molecular proportions

	Solar	Garnet–lherzolite	White collection	Earth model top of mantle
or		6	0.3	3
ab		12	6	10
an	3	29	32	26
di	9	67	26	42
hy		276	202	269
ol	928	611	735	651
le	3			
ne	31			
cs	26			

This notion of allocating the oxides to various sub-systems in an orderly manner is now to be taken one step further and applied globally.

IV. POOL MODEL

Consider a container or pool, as illustrated in Fig. 6.1, holding a number of distinct constituents which are progressively withdrawn or extracted and deposited in a second container, the receiver. The constituents here are taken to be oxide molecules (but could be apples, pears, plums or any other distinct objects). In the pool, let there be n such constituents of number x_i: $i = 1, n$

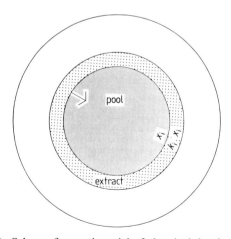

Fig. 6.1. Schema for pool model of chemical development.

(and masses m_i). In the extract, let the number of the constituents be $k_i x_i$, where the k_i are given ratios, the molecular fractionation factors. If the total number of the extract placed in the receiver is t, then for some ξ, conservation of each of the constituents requires:

pool constituent change $dx_i = -\xi k_i x_i$,
total receiver change $dt = -\sum dx_i = \xi \sum k_i x_i = \xi\omega$, say, where $\omega = \sum k_i x_i$, and hence $dx_i/dt = -k_i x_i/\omega$.

These n equations for x_i can then be integrated given $x_i = x_i(0)$ at $t = 0$, to give the pool numbers $x_i(t)$: the integration proceeding until at most all $x_i = 0$ when the pool is exhausted.

Clearly the behaviour depends on the ratios of the k_i. If we wished, the k_i could be "normalized" by, say, setting $\sum k_i = 1$. It is, however, often simpler to think about a particular situation by setting a particular $k_i = 1$.

There is a minor difficulty not immediately apparent. With the proposed extraction strategy the amount of some constituents in the pool becomes exhausted. It is therefore necessary to impose the condition that the $x_i \geqslant 0$. This is straightforward to do during the calculation, since if the next extract would appear to produce an $x_i < 0$ then we set $x_i \equiv 0$. Note that this powerful non-linear effect progressively reduces the "order" (the number of distinct constituents) of the problem. In the extreme case, only one distinct constituent will remain in the pool. Furthermore, ω is a monotonically decreasing function of t and under some circumstances it can reach zero. From this point (namely $\omega = 0$) no further development of the system is possible. Such a state is unrealistic and the corresponding model would be a reject. We therefore have the subsidiary conditions $x_i \geqslant 0$ and $\omega \geqslant 0$.

The amounts x_i diminish monotonically but in different proportions, so it should be obvious that the relative proportions of the constituents are continually varying functions of t and that their proportions are different from their initial values—except for all k_i the same.

(Clearly a more realistic model would allow the k_i to be functions of the thermodynamic state of the material at the core–mantle interface. This would be of considerable interest, especially as there would be a further class of non-linearities arising from the numerous possible phase boundaries available. There is, however, not enough information to allow any sort of calibration of such a model.)

In relating this to a particular planet the information with which we compare the model is usually available as mass rather than numbers of molecules. Thus the mass increment to the receiver is $\xi \sum m_i k_i x_i$, the total mass of the system is $M = \sum m_i x_i(0)$ so that the mass fraction added to the receiver

$$dm = \xi \sum m_i k_i x_i / M$$

IV. POOL MODEL

Hence

$$dx_i/dm = -k_i x_i M / \sum m_i k_i x_i$$

with the $x_i = x_i(0)$ at $m = 0$, and the integration proceeds until the receiver mass fraction $m = 1$, when all the material is in the receiver. The total pool mass fraction is $(1 - m)$. Note that the number dx_i lost to the pool is the number that is in the extract gained by the receiver.

A. Model Behaviour

Since the constituents of the system of interest are dominated by f, m and s the model is essentially a three-component one.

Consider the case $f = m = s$ first with $k(f) = k$, $k(m) = 1$ and $k(s) = 1$. Typical results are shown in Fig. 6.2a. Notice: (1) for k small the pool is depleted in all but f for values of $m < 1$; (2) for k large the pool is depleted in f for values of $m < 1$.

Thus consider next $k(s) = 1$ and find those combinations of $k(f)$ and $k(m)$ such that the pool is depleted in all but f at some value of $m < 1$. Typical results are shown in Fig. 6.2b—the pool is so depleted to the right of the labelled lines.

If we consider the present Earth's pool as not completely depleted in m or s and use this as a criterion with the present mantle mass ratio of 0.673 then we must have $k(f) \geqslant 0.1$ and $k(m) \geqslant 1$. This result is compatible with the estimates of the k_i from the ultrabasic rock data.

The dominant fractionation factor is that for $k(f) = 0.1$–0.2: this leads to an f-enriched core. This is enhanced with $k(m) > 1$ with m depleted with depth.

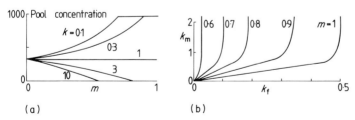

Fig. 6.2. Simple illustrative fractionation model behaviour. (a) Concentration of component 1 in pool as a function of extract mass ratio m for various values of fractionation ratio k. (b) Occurrence of complete depletion of m in the pool, lines of constant extract mass ratio m as a function of fractionation ratios k_f and k_m. Depletion is complete on and to the right of each line.

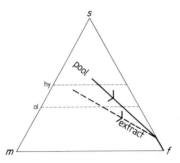

Fig. 6.3. Fractionation paths on (m, f, s) diagram.

The simplicity of the system behaviour is illustrated in Fig. 6.3. This is the usual triangular diagram, here with the three variables f, m and s such that $f + m + s = 1$. The paths of the pool and the extract are both nearly straight lines converging to a point on the side $m = 0$. They pass over the dotted lines representing normative hypersthene $(m, f)s$ and normative olivine $(m, f)_2 s$ to terminate in a phase composed of f with minor s.

B. Preliminary Application

The chemical evolution of the Earth as represented by a particular model is illustrated in Fig. 6.4.

1. The pool (core). Depletion of m and s and enrichment of f proceeds steadily as the core diminishes until the pool is entirely f.
2. The extract (mantle). The concentration of major constituents varies slowly through the early mantle. Beyond $m \approx 0.5$, however, there is a rapid fall in m and s and an increasingly rapid rise in f until the ultimate extract is simply frozen near-pool material.

(Included are the data when an inner core is present. This is referred to below.)

It is important to notice that the extract mass fraction m increases from zero to unity during the development of the chemical structure. This increase is related explicitly to time in Chapter 8. In this and later diagrams illustrating the chemical development, m is used as the independent variable. For example, as shown in Fig. 6.4: the extract silica fraction $s = 0.4$ at the start, with $m = 0$; during the development this fraction rises a little; then it falls sharply to reach $s = 0$ at $m = 0.8$; and thereafter the extract is devoid of silica. The diagrams do not show the distribution of constituents as a function of depth at a particular time—they do show the composition of

V. THE DENSITY DISTRIBUTION

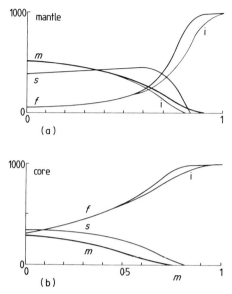

Fig. 6.4. Illustrative example: Earth model chemical development as a function of extract mass ratio m. Values of concentration of m, f and s in: (a) mantle; (b) core. Data for model with and without an inner core.

the core liquid and of the extract (added on to the base of the growing mantle) as a function of the extract mass fraction.

V. THE DENSITY DISTRIBUTION

This pool model provides a sequence of chemical structures. The particular sequence has been chosen to fit the measured estimates of the initial fractionation ratios. Otherwise the model is uncalibrated. We can, however, from the chemical structure obtain the density distribution and compare this with that obtained by global seismology.

There is a complication, namely the role of compressibility of rock substance. For the measured density distribution the effect of compressibility can be removed to obtain the "zero pressure" density, ρ_0, as a function of depth in the Earth. This is the density of material which is decompressed but in which the internal atomic structure is unchanged. Thus ρ_0 is not necessarily the density of a lump of rock substance if it were transported to the surface. (These matters are discussed in Chapter 7. The reader unfamiliar with this could with benefit make a brief detour to that chapter.)

Thus we estimate the zero pressure density distribution for the chemical sequence of the pool model, and compare with the measured values. Not surprisingly there is a further complication, namely how to estimate the density of a given rock substance. Strictly, a rock substance is composed of an ensemble of distinct minerals and the only information we have here is the gross chemical composition. Our knowledge is as yet inadequate to confidently predict the corresponding mineral distribution, the mineral structure or the mineral density. Recent laboratory experiments at pressures typical of those within the Earth, of order 1 Mbar, do however suggest that many minerals at sufficiently high pressure take up a common simple structure. This suggests that a density model in which atoms, represented as midget billiard balls, are more or less close packed will be a fair first approximation. This is what is done here.

A. Density of Rock Substance

Consider a rock substance specified by its n oxides, $\{u_i\} = (f, m, c, n, k, a, s)$, for $i = 1, n$, where $u = X_j O_k$. For example, in this notation $a = Al_2 O_3$ has $X = Al$, $j = 2$, $k = 3$. Let the oxide molecular proportions be $\{q_i\}$. Then the number of oxygen and metal atoms is respectively

$$s_0 = \sum_1^n k_i q_i \qquad s_j = j_i q_i$$

(The atomic fraction of oxygen is $s_0/(s_0 + \sum s_i)$.)

Let the metal atoms have mass $M_i m_H$ and ionic radius a_i, where $m_H = 1$ amu (atomic mass unit, 1.66×10^{-27} kg). Similarly for the oxygen take mass $M_0 m_H$ and ionic radius a_0. Then the total mass and volume of the atoms is

$$M = m_H \sum_0^n s_i M_i \qquad V' = \tfrac{4}{3}\pi \sum_0^n s_i a_i^3$$

Let the packing fraction $p' = V'/V$, where V is the total volume (atoms + interstitial voids) and V' is the volume of the atoms alone. Then the density of the substance is $\rho = p' M/V'$.

This expression can be put more usefully by taking oxygen as the reference material. Thus write $M_i' = M_i/M_0$, the atomic weight relative to that of oxygen, so that (Fe', Mg', Ca', Na', K', Al', Si') = (3.48, 1.52, 2.51, 1.44, 2.44, 1.69, 1.75); $a_i' = a_i/a_0$, the ratio of ionic radius to that of oxygen, so that $M_0' = 1$, $s_0' = 1$, $a_0' = 1$; and $s_i' = s_i/s_0$, the ratio of numbers of a particular atom to the number of oxygens. Also introduce the matter density of the oxygen ion, namely $\rho_* = 3 M_0 m_H / 4\pi a_0^3$.

Then we have

$$\rho = p' \rho_* \sum_0^n s_i' M_i' \bigg/ \sum_0^n s_i' a_i'^3$$

V. THE DENSITY DISTRIBUTION

Note that if we measure ionic radii in angstroms, $\text{Å} = 10^{-10}$ m, we have $\rho_* = \xi M_0/(a_0/\text{Å})^3$ with $\xi = 3m_H/4\pi\text{Å}^3 = 396.4 \text{ kg/m}^3$, so that with $a_0 = 1.4 \text{ Å}$ the matter density of the oxygen ion is 2310 kg/m^3.

There are a host of possible structures of possible interest. There is, however, one very powerful point to note. A mineral is to a first approximation simply a "framework" of oxygen atoms into which the metal ions are inserted. The oxygen makes up the bulk of the ionic volume since it is numerically the most abundant ion and usually the largest. The volumetric ratios for the ions in the rock substance derived from the solar abundances are shown in Table 6.5—oxygen makes up 93.3% of the volume, the minor

Table 6.5 Ionic data for rock substance constituents

	a (Å)	n (ions)	Volume ratio	Molecular mass as oxide
O^{2-}	1.40	1458	0.933	
Fe^{2+}	0.74	316	0.030	71.9
Mg^{2+}	0.66	296	0.020	40.3
Ca^{2+}	0.99	21	0.005	56.1
Na^+	0.97	20	0.004	62.0
K^+	1.33	2	0.001	94.2
Al^{2+}	0.51	24	0.001	101.9
Si^{4+}	0.42	345	0.006	60.1

In this work all Fe as Fe^{2+}.
Ionic radii data originally by L. H. Ahrens as discussed in Shannon and Prewitt (1969).

constituents Ca, Na, K and Al only 1%, and the major positive ions Fe, Mg and Si 5.6%. It is one of those minor (?) miracles of nature that the amounts of the common atoms have this nice relationship. Let us therefore take the packing factor p of the oxygen ions alone, in which case $V' = \frac{4}{3}\pi s_0 a_0^3$, and the net density

$$\rho = p\rho_* \sum_0^n s'_i M'_i$$

B. The Critical Level

The atomic structure of a substance is a function of its thermodynamic state. With increasing pressure a particular mineral undergoes packing changes which generally lead to denser, more closely packed forms. A rock substance, being an ensemble of mineral phases, follows a similar pattern, although the effect is somewhat blurred since the different minerals change at different states. Nevertheless, in the Earth's mantle we find a rather distinct density change at a pressure of 0.15 Mbar, radius ratio about 0.935, mass ratio about 0.87, from zero pressure density of about 3300 to 4000 kg/m^3 (for details refer to Chapter 7). I shall refer to this as the "critical (packing) level" and take it to be at the same pressure in each of the terrestrial planets. In the models the change is taken as a discontinuity and in crossing it the packing factor is changed abruptly.

This effect does not substantially alter the results for the larger terrestrial planets since it affects a relatively small part of the total mass, but for the smaller terrestrial planets it is powerful. Note especially that for the Moon the critical level is not reached so that in the Moon model the packing remains at the low pressure value.

C. The Packing Factor

The density model, then, is determined by the choice of the packing factor p. There is a range of possible values, among which typically are the following:

1. $p = \pi/6 = 0.524$, cubical packing;
2. $p = \pi/3\sqrt{3} = 0.605$, two-dimensional closely packed sheets;
3. $p \approx 0.637$, random packing;
4. $p = 2\pi/9 = 0.698$, closely packed sheets sitting proud on one another, giving a stiff framework—not far from the spinel structure with $p \approx 0.70$;
5. $p = \pi/3\sqrt{2} = 0.7405$, three-dimensional close packing with a more or less rigid framework.

Somewhat arbitrarily (after considerable detailed work) I have chosen p as follows. For low pressures, random packing gives a fair fit to the zero pressure density ρ_0 of the upper mantle rocks of interest, hence $p = 0.637$. This is the same in principle as the assumption of global seismology that the density of the top of the mantle is 3300 kg/m^3. For high pressures, the choice of packing similar to that of the spinel structure gives values of $\rho_0 \approx$ 4000 kg/m^3 in the deep mantle and $\rho_0 \approx 6600$ kg/m^3 in the core of the Earth as measured (see Chapter 7), hence $p = 0.70$.

D. The Mantle–Core Density Contrast

The most striking feature of the measured Earth's density distribution is the strong density contrast at the core–mantle boundary. This is an intrinsic

VI. SIMPLE MODEL WITH UNIFORM PACKING 121

feature of the model presented here. It is represented by the ratio $\xi = \rho_{0(core)}/\rho_{0(mantle)}$ evaluated at the mantle–core interface.

VI. SIMPLE MODEL WITH UNIFORM PACKING

Before going on to a more elaborate version of the model, consider a model, appropriate to the Earth, in which the packing factor is the same throughout the entire body. Let us choose uniform random packing $p = 0.637$ in both mantle and core. The results are shown in Fig. 6.5.

1. The mantle ρ_0 rises very slowly near $m = 0$. This behaviour is that found in the actual Earth.
2. Beyond a particular value of m the mantle density ρ_0 rises rapidly.
3. The core density ρ_0 rises monotonically until in this particular model the core is depleted in all except one constituent—after which both mantle and core materials are the same.
4. The zero pressure density ratio ξ at the core–mantle interface rises from 1.19 to a peak value of 1.45 at $m \approx 0.65$ and falls to unity at and after $m \approx 0.87$.
5. The atomic fraction of oxygen in the mantle is nearly constant until it falls in the same manner as ξ, whereas in the core the ratio falls monotonically to unity at $m \approx 0.8$.

Overall this is just the pattern we have in the actual Earth—although there is no direct evidence on the oxygen distribution. There are, however, a number of features which do not fit the Earth at all well.

1. The zero pressure density at the top of the mantle is just what it should be—of course, that is built into the model but the density in the lower mantle is too small. Clearly closer packing is required and indeed in the full models the packing is changed to $p = 0.7$ below the critical level. Similarly the core liquid densities are too low. In the full model I have therefore just taken $p = 0.7$ in the core throughout the development. Strictly a packing change will occur but certainly in the larger bodies the bulk of the core mass will be at pressures greater than that of the critical level. In the case of the Moon, however, where the critical level is not reached at all, I retain $p = 0.637$.
2. At some value of mantle mass, less than unity, $\xi = 1$. The pool is depleted in all but f. This is quite unrealistic for this model and we now consider the consequences. (In reality the terrestrial planets would never reach this point anyway—since, as we shall see in Chapter 8, there is not enough time.)

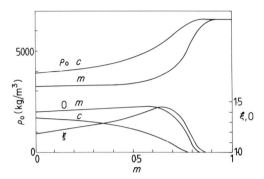

Fig. 6.5. Illustrative example: Earth model zero pressure density development as a function of extract mass ratio m. Values for mantle, core and density ratio ξ. The oxygen atomic ratio in the mantle and core are also shown.

A. The Central Chemically Homogeneous Mass

The functions $\rho_{0(\text{mantle})}$ and $\rho_{0(\text{core})}$ are functions of m and both tend to a common central value since the pool is then depleted in all but iron oxide. Over a range of m near unity they are indistinguishable. It is of interest to specify when this occurs. Noting that in practice the actual density distribution in the Earth as found by global seismology is at best determined to $\pm 1\%$, let me define the "central mass" as that portion of the body for which the mantle and core zero pressure densities differ by 1% or less, namely for which

$$(\rho_{0(\text{mantle})} - \rho_{0(\text{core})})/\rho_{0(\text{core})} \leqslant 0.01$$

The "central mass fraction" is shown in Fig. 6.6 as a function of f (as is $\bar{\rho}_0$ for various values of the critical level). The central mass fraction is independent of the critical level and as expected increases with f.

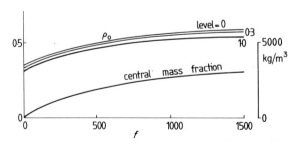

Fig. 6.6. Global mean zero pressure density as a function of concentration f for various critical levels. The central mass fraction is also shown.

VII. THE INNER CORE   123

For the bodies of interest here with f no more than 500, the central body is quite small, of mass no more than 0.17 of the total mass. The central body would become a more or less distinct object late in the geological life of a planet, although for the moment I leave aside how this would come about.

B. On the Choice of f

There is a further constraint we must now relax. So far we have treated the terrestrial planets as if they were chemically identical. Plainly this is not so. The mean zero pressure density $\bar{\rho}_0$ differs from planet to planet, from 3300 kg/m^3 for the Moon to 5100 kg/m^3 for Mercury. (Note that the role of compressibility and packing changes has effectively been removed from this data.) The only major quantities we have not so far allowed to vary are the global chemical proportions. Rather than letting them all take a variety of values, let us consider the dominant chemical ratio f/m. If we take m at its nominal value we have a class of models which is a function of f. In other words we have $\bar{\rho}_0 = \bar{\rho}_0(f)$. At this point the approach is *ad hoc* and we will need to consider later whether or not there is any evidence to support our choice.

Thus we choose f for each planet so that the model $\bar{\rho}_0$ is reasonable, and in addition for the Earth obtain reasonable values for $\bar{\rho}_{0(\text{mantle})}$ and $\rho_{0(\text{core})}$.

VII. THE INNER CORE

While the pool is sufficiently hot and rich in the material able to produce the low-density mantle, the sole extract from the pool is deposited at the mantle–pool boundary. Otherwise, some minerals may form which are sufficiently dense to fall inwards and the pool will be insufficiently hot to remelt them before they accumulate near the centre of the pool. Under these circumstances an inner core forms. (For the present Earth the inner core is of mass fraction 0.023, with Poisson's ratio close to 0.5, the value for a liquid, and is probably a cumulate mush; see Chapter 7.) The pool is then depleted by two extraction processes, that at the mantle–pool boundary and that at the inner core–pool boundary.

The formalism of the pool model can be applied as it stands. With the fractionation ratios rewritten as k'_i for the mantle material and k''_i for the inner core material the net fractionation ratios are $k_i = k'_i + k''_i$ with the number proportions $k'_i x_i$ in the mantle extract and $k''_i x_i$ in the inner core extract.

A number of simplifications can be made.

1. The mantle is built from the lighter m-rich fraction so that the pool becomes enriched in a single major component f. Let us therefore make the strong assumption that the inner core is composed solely of f. Thus we take all the $k_i'' \equiv 0$ except for $k_1'' \equiv k_{\text{(inner)}} > 0$.
2. The pool contains the major components (m, f and s). With the choice of k_i' here, m and s are depleted in the pool very roughly in similar proportions. The composition of the pool is then largely specified by stating the ratio $\mu = m/(m+f)$. Let us therefore assume that the inner core will begin to form when $\mu < \mu_c$, where μ_c is some critical value. Strictly, of course, the onset of the inner core will be determined not only by the chemical proportions but also by the thermodynamic state of the pool a nice self-consistent field problem, which one must leave to the future.

It remains, then, to complete the specification of the model, to estimate $k_{\text{(inner)}}$ and μ. It is very fortunate that for the present Earth we have measured

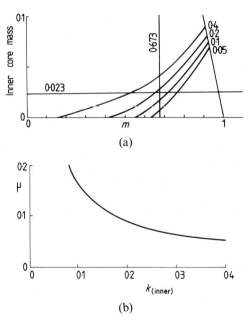

Fig. 6.7. Choice of inner core fractionation factors for the Earth. (a) Inner core mass fraction as a function of extract mass ratio m for various values of $\mu = m/(m+f)$ with $k_{\text{(inner)}} = 0.1$. (b) Core composition ratio μ as a function of $k_{\text{(inner)}}$ to obtain an inner core of mass fraction of 0.023.

VIII. MODEL RESULTS FOR TERRESTRIAL PLANETS

values of both the core mass ratio, 0.327, and the inner core mass ratio, 0.023. Thus as shown in Fig. 6.7a for a given value of $k_{(inner)}$ we obtain, for different values of μ, different curves for the inner core mass as a function of mantle mass m using the pool model. And as shown in the example for $k_{(inner)} = 0.1$ the curve for $\mu_c = 0.165$ gives a fit to the Earth data. In Fig. 6.7b the relation between μ_c and $k_{(inner)}$ is shown for the Earth. There is no obvious choice of the most suitable pair of values, though very small values of $k_{(inner)}$ or μ_c would seem unlikely. I choose $k_{(inner)} = 0.1$ and $\mu_c = 0.165$.

The inner core is included in the model results to be described below. The major items are summarized in Table 6.6. Although the inner core contains a small part of a planet's mass it has a profound effect on the later stages of the planet's development.

Table 6.6 Inner core, model results

	$m_{(inner)}$ [a]	$m_{(inner)(max)}$ [b]	ξ final	$m_{(inner\ core)}$ now
Mercury	0.24	0.133	1.03	0.08
Venus	0.57	0.064	1.04	0.024
Earth	0.50	0.078	1.04	0.023
Moon	0.86	0.010	1.15	0
Mars	0.68	0.040	1.02	0.002

[a] Mass of mantle at which inner core starts to form.
[b] Final mass of inner core.

Thus our naïve concept of a "central mass" has been replaced by that of an inner core. We note that, apart from the Moon and perhaps Mars, all the terrestrial planets will have an inner core now. There remains a small density contrast at the final mantle–core boundary, when the base of the mantle meets the top of the inner core. The final pool fluid would be enhanced in the originally minor constituents which have fractionation coefficients less than unity to produce an unusual "pegmatitic" rock substance.

VIII. MODEL RESULTS FOR THE TERRESTRIAL PLANETS

The model parameters chosen are given in Table 6.7, and the results are shown in Figs 6.8–6.12 and Table 6.8.

Mercury. This is the "olivine" planet with dominant normative olivine in the mantle until $m \geqslant 0.65$. The relatively large inner core is a prominent feature.

Table 6.7. Chemical history model parameters

	$\bar{\rho}$ (kg/m³)	Critical depth (mass fraction)	f	Final inner core (mass fraction)
Mercury	5420	0.326	500	0.22
Venus	5250	0.143	250	0.12
Earth	5520	0.126	320	0.13
Moon	3340	—	50	0.02
Mars	3940	0.331	80	0.09

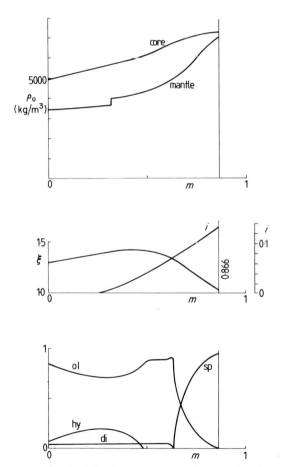

Fig. 6.8. Mercury: chemical development properties as a function of mantle mass fraction m. (a) Zero pressure density ρ_0 for mantle and core. (b) Density ratio ξ and inner core mass fraction. (c) Normative mineral distribution in mantle.

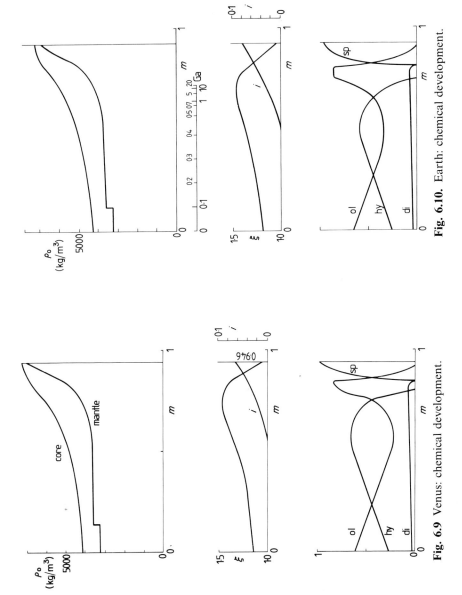

Fig. 6.10. Earth: chemical development.

Fig. 6.9 Venus: chemical development.

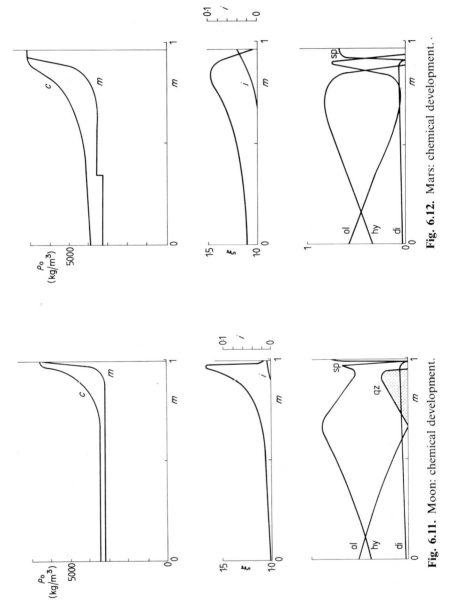

Fig. 6.11. Moon: chemical development.

Fig. 6.12. Mars: chemical development.

VIII. MODEL RESULTS FOR TERRESTRIAL PLANETS

Table 6.8. Normative mineral assemblage near the top of the mantle (at depth 0.02a—e.g. for Earth at depth 130 km): molecular proportions

	or	ab	an	di	hy	ol
Mercury	3	10	26	42	84	835
Venus	3	10	26	42	303	617
Earth	3	10	26	42	269	651
Moon	3	10	26	42	400	519
Mars	3	10	26	42	351	568

The value of $\bar{\rho}_0$ for Mercury cannot be independently checked: using the data of Fig. 6.6, with critical level 0.326, allows values of f as high as 750 for no compression at all, or as low as 320 for compression as great as on Earth. Both these limits are unrealistic extremes. Values of f of 400–600 would be possible with values of $\bar{\rho}_0$ of 4950–5250 kg/m³. The amount of f in Mercury, then, is enhanced over the solar abundance by about 50%.

Venus. Venus and Earth are "twin sister and brother" with quite minor differences.

Earth. This is, of course, our reference planet, the only terrestrial planet for which we can have much confidence in our knowledge of the interior. It should be clear that the other terrestrial planets are modelled in Earth's image solely because we do not have enough information to do any better. Such information as exists suggests that this is a good approximation but it should not be a future surprise to find it is not so.

The bulk of the mantle forms as a normative olivine–hypersthene body, making a peridotite/pyroxenite rock. There is a pronounced peak in olivine at a later stage when the pool silica is becoming exhausted.

This is the only body for which we have a direct measure of the global mean zero pressure density $\bar{\rho}_0$. From the work described in Chapter 7, $\bar{\rho}_0 = 4700 \pm 100$ kg/m³. The fit is reasonably good to present values: (1) $\bar{\rho}_0$; (2) $\rho_{0(mantle)} = 4000$ kg/m³ and $\rho_{0(core)} = 6600$ kg/m³. This is using the solar value of f. The sensitivity to f has been shown in Fig. 6.6: $\bar{\rho}_0$ between 4700 and 4900 kg/m³ requires f between 280 and 350. This suggests that for the Earth, in terms of this model, $f = 320 \pm 10\%$.

If we try to refine the fit by varying the fractionation factors we find that over the acceptable range of these factors there is only a marginal change in $\bar{\rho}_0$. For $k_1 = 0.1$ to 0.2 and $k_2 = 1$ to 2 we have $\bar{\rho}_0 = 4820$ to 4940 kg/m³.

A time scale, obtained from the work of Chapter 8, is included on the diagram. What is striking is the rapid slowing down of the development after 1 Ga. During all of the currently observable geological time, now to

4 Ga ago, the change in the gross features of the Earth's interior have been very small.

Moon. This is a small, relatively simple body. The zero pressure densities of the mantle and core are nearly the same and nearly constant until the final stage of development. The mantle development is dominated by normative hypersthene and is in this respect a body similar to Mars.

For bodies of small mass and low density, notably the Moon, compressional effects on mean density are small so that $\bar{\rho}_0 \approx \bar{\rho}$. This allows limits to be set on f directly. With $\bar{\rho} = 3340 \text{ kg/m}^3$ the relation $\bar{\rho}_0(f)$ requires $f \approx 50$.

The Moon is chemically most unusual. Whereas all the other bodies are entirely undersaturated in silica, a normative quartz phase is present in the Moon model from $m = 0.67$ to 0.96, together with normative hypersthene making a quartz pyroxenite, an unusual and unexpected rock. (Whether or not one should take this possibility seriously I leave to the reader.)

Mars. This is the "pyroxene" planet with dominant normative hypersthene in the mantle until $m \geqslant 0.9$. This is also a body of small mass with $\bar{\rho}_0 \approx \bar{\rho}$,

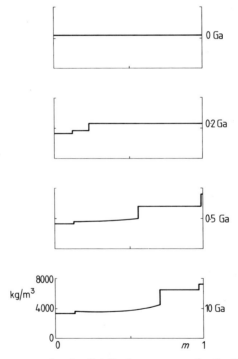

Fig. 6.13. Zero pressure density distribution sequence for the Earth as a function of mantle mass m at various nominal times 0, 0.2, 0.5 and 10 Ga.

so that from the measured $\bar{\rho}$ we have directly $f < 90$. A nominal value of $f = 80$ is used here.

IX. THE DENSITY DISTRIBUTION SEQUENCE

The point of view of this chapter is summarized in Fig. 6.13, which shows the profiles of ρ_0 at various stages of the sequence of the structural development of a terrestrial planet. The times are obtained in Chapter 8.

1. $t = 0$. This is the starting point of the pool model, an entirely liquid body.
2. $t \sim 0.2$ Ga. The mantle is already developed to depths below the critical level. The core density is already enhanced following the removal of lighter material to form the mantle.
3. $t \sim 0.5$ Ga. The mantle is well developed. The inner core is established.
4. $t \sim 10$ GA. The body is in an advanced state of development with 70% of the body mass in the mantle. The inner core is a small but prominent feature. (This is as far as we go, since at about this time the Sun blows up into its sub-giant stage and all the planets will be enveloped and disintegrate.)

CHAPTER 7

Physical Development

I. INTRODUCTION

The physical structure of a planet of given mass is determined by the type and distribution of the kinds of matter of which it is composed, together with the effects of compressibility of rock substance in its own gravitational field. The physical structural development is now considered as an ordered sequence of "snapshots" as yet unrelated to time. For each planet the sequence covers the development from completely liquid to completely solid following the corresponding chemical development described in Chapter 6 and provides the particular paths through the structural "phase-space" of the class of all available models of the type considered here.

The basis of all this work is our knowledge of the Earth's interior obtained from global seismology. We look first at this classic piece of geophysics and then apply a structural model of this type to the terrestrial planets.

The seismic model data can be used to calibrate an equation of state. It is this equation of state which can then be used as the basis of models of physical structure of the terrestrial planets now and in the past. The novel feature, presented in this work, is the use in the equation of state of the zero pressure density obtained from the chemical structure, described in Chapter 6.

Although the available data is inadequate to satisfactorily calibrate similar structural models for the other non-gaseous bodies of the solar system, some comments on the structure of the moons of the Jovian planets are made here, too.

II. EARTH PHYSICAL STRUCTURE NOW

The present radial structure of the mechanical properties of the Earth, obtained from the travel times of seismic waves or from the frequencies of elastic vibrations of the whole Earth, averaged over spherical surfaces, is known to about $\pm 1\%$ or so. In this work I take the seismic model data as given (details can be found in a number of excellent books, e.g. Bullen, 1975; Cook, 1980; Jeffreys, 1976).

In order to use this knowledge of the Earth's interior to construct models of the other terrestrial planets, and to construct models for the terrestrial planets as they were in the past, I will use the Earth data to calibrate a manageable equation of state for rock substance. This equation of state can then be used in a model of a quasi-hydrostatic body to represent the structure.

Furthermore, in order to identify the substance which constitutes a particular zone within the Earth it is most straightforward to work with the density at zero pressure obtained by removing solely the effects of compressibility.

The data on which the discussion here is based is taken from the revised Model A of Bullen and Haddon (1967), quoted in Jeffreys (1976, p. 213). This particular model is calibrated using the value $C/Ma^2 = 0.3308$. The data are shown in Table 7.1. (I have interpolated the data, using polynomials of order 3 or more in the radius, for the core to fractions of the external radius rather than as quoted.)

A. The Equation of State

A variety of empirical relations have been used for an equation of state. A particularly convenient one is obtained by noticing that the compressibility $\chi = (\partial \rho/\partial p)/\rho$, in a particular layer, closely satisfies

$$\chi \rho^n = \chi_0 \rho_0^n = C \text{ (a constant)} \qquad \rho = \rho_0(1 + n\chi_0 p)^{1/n}$$

The form of the density as a function of pressure for this relation is shown in Fig. 7.1 for the materials of the outer core and the lower mantle of the Earth.

For each of the layers of the Jeffreys model (usually referred to as layers A–G, crust–inner core) the equation of state can be fitted to the data by choosing appropriate values of n, ρ_0 and C to give a minimum root-mean-square difference between the equation and the data.

(The units of C are awkward—$(kg/m^3)^n/Pa$. Values of C are stated here solely by value, in these SI units.)

Table 7.1. Radial distribution in the Earth as a function of dimensionless radius r

	r	m	ρ (kg/m^3)	p (Mbar)	k (Mbar)	σ
B	1.00	1.000	3320	0.00	1.16	0.270
	0.98	0.965	3440	0.05	1.33	0.275
	0.96	0.929	3540	0.10	1.52	0.277
	0.94	0.893	3640	0.14	1.74	0.280
C	0.92	0.959	3940	0.19	2.30	0.282
	0.90	0.822	4190	0.24	2.75	0.278
	0.88	0.786	4370	0.29	3.09	0.275
	0.86	0.750	4500	0.35	3.36	0.274
	0.84	0.714	4590	0.41	3.56	0.276
D	0.82	0.680	4660	0.47	3.74	0.277
	0.80	0.646	4730	0.53	3.93	0.279
	0.78	0.613	4800	0.59	4.13	0.281
	0.76	0.583	4870	0.65	4.33	0.284
	0.74	0.553	4940	0.71	4.54	0.286
	0.72	0.525	5010	0.77	4.76	0.289
	0.70	0.497	5080	0.84	4.98	0.292
	0.68	0.470	5140	0.90	5.20	0.294
	0.66	0.446	5200	0.97	5.41	0.296
	0.64	0.422	5270	1.03	5.61	0.297
	0.62	0.400	5330	1.10	5.84	0.299
	0.60	0.378	5390	1.17	6.04	0.300
	0.58	0.357	5460	1.24	6.26	0.301
	0.56	0.337	5520	1.31	6.33	0.298
	0.548	0.327	5560	1.36	6.39	0.299
E	0.548	0.327	9980	1.36	6.55	0.50
	0.50	0.255	10,450	1.69	7.49	0.50
	0.45	0.190	10,880	2.01	8.68	0.50
	0.40	0.136	11,250	2.31	9.76	0.50
	0.35	0.092	11,570	2.60	10.78	0.50
	0.30	0.059	11,840	2.86	11.71	0.50
	0.25	0.034	12,060	3.08	12.66	0.50
F	0.20	0.018	12,230	3.26	13.63	0.45
	0.15	0.008	12,370	3.43	14.31	0.45
	0.10	0.002	12,444	3.54	14.37	0.45
	0.05	0.000	12,480	3.60	14.44	0.45
	0.00	0.000	12,510	3.62	14.50	0.45

m = mass ratio, ρ = density, p = pressure, k = bulk modulus, σ = Poisson's ratio.
Adapted from Bullen and Haddon (1967), their Model A.

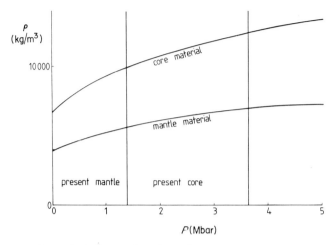

Fig. 7.1. Density ρ (kg/m^3) as a function of pressure P (Mbar) for present Earth lower mantle and core rock substance.

The identification of the zero pressure density ρ_0 is illustrated for the core in Fig. 7.2. This shows as a function of n the goodness of fit of ρ_0 and of C given as rms of the proportional variation for the ten data points available. The corresponding ρ_0 is also shown. A number of features are of note.

1. The goodness of fit in C has a distinct minimum with a goodness of fit of 1.4%.
2. The variation of ρ_0 (over the data points, for a given n) is monotonic, small near $n = 3$ and rising to large values beyond $n = 4$. It would not be possible to choose a "best" value of ρ_0 just from this variation.
3. The density ρ_0 varies monotonically and has the value near the best fit (minimum variation is C) of $\rho_0 \approx 6600$ kg/m^3.
4. The high precision of the fit should not fool us into considering the result as highly accurate. There are many arbitrary features of the whole seismological model. It is not possible to put satisfactory limits on ρ_0—it could be ± 200 kg/m^3 or more.
5. The calculations are not just an exercise in empiricism. The extrapolation of ρ to zero P is obtained from the relation $\rho(P)$ formed from the system itself. This is certainly likely to be a better method of estimating than using, say, a theoretical model of $\rho(P)$.
6. It is of interest to note that ρ_0 in the inner core is not significantly different from the value elsewhere in the core.

II. EARTH PHYSICAL STRUCTURE NOW

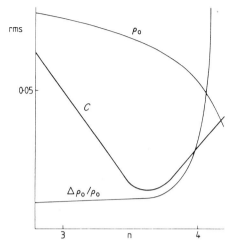

Fig. 7.2. Identification of parameters of present Earth core material. Root mean square (rms) of fit to Jeffreys–Bullen core data as a function of compressibility parameter n for C, $\Delta\rho_0/\rho_0$. Also shown, ρ_0 (10^5 kg/m^3).

B. Radial Variation of ρ_0

The radial variation of zero pressure density $\rho_0(r)$ is shown in Fig. 7.3 and the seismic layer structure in Table 7.2. What is immediately striking is that once the simple effect of compression is removed we see that the Earth is made of zones within which the material is (radially) homogeneous. This is

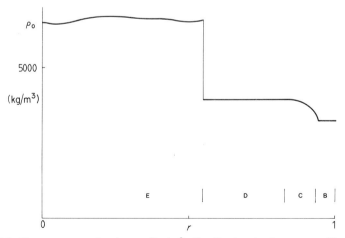

Fig. 7.3. Zero pressure density ρ_0 (kg/m^3) distribution in the present Earth as a function of dimensionless radial coordinate r.

Table 7.2. Density parameters in major layers of the Earth

Layer	r	P (Mbar)	m	n	ρ_0 (kg/m^3)	C	rms %
A	–	0.009	0.004	–	2840	–	–
B	0.935	0.141	0.111	3.00	3300	0.298	5.9
C	0.83	0.408	0.206	2.91	3470–3990	0.127	6.3
D	0.548	1.361	0.352	3.16	3980	1.036	1.2
E+F+G	0	3.62	0.327	3.65	6610	619	1.4

1. A = crust; B + C + D = mantle; E + F + G = core; layer A, nominal thickness 33 km.
2. r = radius, P = pressure at base of layer; m = mass fraction of layer—inner core mass fraction 0.023.
3. rms = root mean square of C in layer.
4. Fitted values of C are insensitive to n in layers B and C.
5. Units of C are (kg/m^3)^n/Pa, namely with density in kg/m^3 and pressure in Pa.

so except for zone C—which in any event is the most arbitrary of the seismological model.

III. THE MODEL RELATIONS

Consider a radially symmetric self-gravitating body composed of a set of k shells, labelled $i = 1, k$, each of which is chemically homogeneous and characterized by an equation of state with given (ρ_{0i}, n_i, C_i). This study is restricted to the case $k = 4$ with layers corresponding to those labelled B, C, D, E + F + G in the Jeffreys–Bullen model. Layers B, C and D of the mantle are made of solid rock substance; layers E, F and G of the core are here treated as a single liquid unit, the inner core being ignored. The crustal layer A is also ignored, layer B being considered as extending to the solid surface.

The structure will be regarded as determined if the radial density distribution $\rho(r)$ is known.

The system is assumed to be in hydrostatic equilibrium. Hence the equations, with: radial coordinate r; mass m enclosed within a sphere of radius r; density ρ; acceleration of gravity g; and pressure P; and quantities of interest derived from the model, the moment of inertia I and gravitational energy W are as follows.

It is convenient to write these relations in partially dimensionless form by writing $m' = m/M$, $r' = r/a$, where a is the external radius (and then dropping the prime):

III. THE MODEL RELATIONS

$$dm/dr = 3\rho r^2/\bar{\rho}$$
$$g = g_s m/r^2 \qquad g_s = GM/a^2$$
$$dP/dr = -ga\rho$$
$$\rho^n = \rho_0^n + nCP$$
$$d(I/I_0)/dm = \tfrac{5}{3}r^2 \qquad I_0 = \tfrac{2}{5}Ma^2$$
$$d(W/W_0)/dm = -\tfrac{5}{3}rg/g_s \qquad W_0 = \tfrac{3}{5}GM^2/a$$

A key quantity used in the identification of the model structures is the moment of inertia factor (or ratio):

$$\xi = I/Ma^2$$

For a homogeneous body $\xi = 0.4$. For inhomogeneous bodies, such as those of interest here, with density increasing with depth, $\xi < 0.4$. (Note that in this work the moment of inertia is taken to be that about the polar axis, usually given the symbol C.)

The boundary conditions are $r = 1$, $m = 1$, $P = 0$ (and $\rho = \rho_{01}$); $r = 0$ and $m = 0$; and at the interface between layers i and j the pressure is continuous so that $P_i = P_j$. Hence in crossing an interface the current value of P is retained but (ρ_{0i}, n_i, C_i) is replaced by (ρ_{0j}, n_j, C_j) and a new density is obtained from the new equation of state—and as a consequence there will be a discontinuity in density. It is important to note the assumption here that the phase transformations occur at particular pressures. The interface pressures used here are those of the Jeffreys–Bullen model: 0.157, 0.429 and 1.375 Mbar.

The essence of the numerical task is to find the size of a body of given mass. This is done here by an iterative method.

There is a wide range of suitable iteration techniques. The method used in this work has been to integrate downwards from $r = 1$ (in steps $dz > 0$, $dr = -dz$) and use the central value of $m = m_c$, usually non-zero, to select a revised radius using the method of bisection of the interval. Thus start with given $a_1 < a_2$ and $a = \tfrac{1}{2}(a_1 + a_2)$. Two cases occur: (1) $m_c > 0$, reset $a_1 = a$; (2) $m_c < 0$ (namely m reaches zero for $r > 0$, and in practice the integration downwards is simply terminated), reset $a_2 = a$. This is repeated until the new iterate a' is such that $|a' - a| < \Delta a$, where Δa is a preset precision, here taken usually as 1 or 0.1 km. In practice, for a given final precision it is quicker to start with large $|a_2 - a_1|$ and Δa and progressively lower Δa to the final precision required.

When a particular model structure as it exists now has been identified, a major interest is then to evaluate the structure as it changes owing to the growth of the mantle. In the model calculations we simply integrate downwards, as before, until the mantle mass ratio reaches a set value, and then force a change to the core material. Thus:

1. While $m_{(mantle)} < m_{(set)}$, proceed normally, at each interface changing from material i to $(i+1)$ using the condition $P_i = P_{i+1}$ in a sequential manner for $i = 1$ to k.
2. When $m_{(mantle)} \geq m_{(set)}$, reset material i to k, regardless of i, and again impose the condition $P_i = P_k$.

In this manner some of the layers B, C and D may not be encountered; indeed for $m_{(mantle)} = 0$ none of them are, since the body is entirely liquid.

IV. METHOD OF IDENTIFICATION OF THE MODELS

The identification of a particular model is made from the measured values of mass, radius and moment of inertia. Thus for a given mass the model structure is varied in a systematic manner until the model and measured values of (a, ξ) agree to within preset limits. We do not, however, have measured (a, ξ) data for all the terrestrial planets; for Mercury and Venus there are no measurements of ξ, and for these bodies other constraints must be used.

There is, however, a (surprising?) correlation $\xi(\bar{\rho})$ indicated in Fig. 7.4, which shows the measured data for the Moon, Mars and the Earth, and the model values for Mercury and Venus proposed below. This correlation gives some confidence in the Mercury and Venus models.

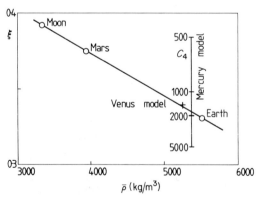

Fig. 7.4. Correlation of moment of inertia ratio ξ with planet mean density $\bar{\rho}$ (kg/m³). Measured values for the Moon, Mars and Earth. Model values for Venus and Mercury. Also shown for Mercury are the values as a function of core compressibility factor C_4.

IV. METHOD OF IDENTIFICATION OF THE MODELS

There is another major lack of information arising from ignorance of the parameters of the equation of state. We have some information about the (zero pressure) reference density from the pool model but no information on n, C. Possibly the values of n, C for the solid mantle phases will be similar to the Earth values, since the chemical composition of these materials as suggested by the pool model are similar. Such is not the case for the core liquid. There are a few scraps of information which at least allow a first attempt at setting up models for the terrestrial planets, but the reader should not take the detailed values too seriously in what is nothing more than a piece of rampant empiricism.

Values of (ρ_0, n, C) use the values of n, C from the Jeffreys–Bullen model data and ρ_0 values from the pool models of Chapter 6. For layer 4, the liquid core, however, values of C_4 are directly obtainable only for the Earth but a good estimate can be made for C_4 for Mars. This gives two points (!) for which an extrapolation/interpolation suggests values for the Moon and Venus. The compressibility parameter C_4 is presumed to be related to the concentration f of iron oxide in the core. The correlation is shown in Fig. 7.5. The C_4 values for Earth and Mars and taken as known, the Venus value

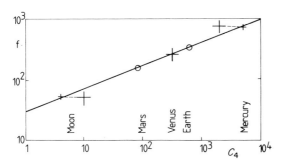

Fig. 7.5. Correlation of concentration f with compressibility factor C_4.

is obtained directly by interpolation, while modified values from those suggested by extrapolation are used for the Moon and Mercury (see below).

Because the scraps of information arise from different aspects of the planets, the sequence of the discussion is in the order: Earth, essentially the Jeffreys–Bullen model with minor modifications to calibrate a strictly 4-layer model; Mars, from which the second C_4 value is obtained; followed by models of dubious validity for Mercury, Venus and the Moon. Table 7.3 lists the parameters used in this work, together with a summary of the results.

Table 7.3. Parameters and model data for the terrestrial planets

	ρ_0 (kg/m³)	n	C	a	ξ	m_{core}	z levels	ρ_c (kg/m³)	P_c (Mbar)
Mercury	3450	3.00	0.298						
	3800	2.91	0.127						
	[a]3800	3.16	1.036						
	5960	3.65	2000	2441	0.3322	0.6	0.260	9900	0.44
Venus	3280	3.00	0.298				0.082		
	3610	2.91	0.127				0.202		
	3880	3.16	1.036				0.350		
	5605	3.65	315	6044	0.3395	0.42		9490	2.43
Earth	3300	3.00	0.298				0.070		
	3700	2.91	0.127				0.172		
	3980	3.16	1.036				0.455		
	6610	3.65	619	6378	0.3312	0.327		12,480	3.56
Moon	3200	3.00	0.298						
	[a]3200	2.91	0.127						
	[a]3200	3.16	1.036						
	3723	3.65	10	1740	0.3920	0.2	0.435	3740	0.05
Mars	3240	3.00	0.298						
	3550	2.91	0.127						
	[a]3550	3.16	1.036						
	4280	3.65	79	3400	0.3752	0.52	0.230	4760	0.30

[a] Level not reached, values irrelevant.

V. IDENTIFICATION

A. Earth Identification

In constructing a model Earth we note, apart from layer C (the lower mantle), that ρ_0 in each layer is more or less a constant. In layer C, however, ρ_0 is 3470–3990 kg/m³. It is convenient to represent layer C also as a layer of constant ρ_0—whereupon we can have a simple 4-layer model. The value of ρ_{03} can be chosen to give a fit to the radius a and the moment of inertia ratio ξ as indicated in Fig. 7.6. A value of $\rho_{03} = 3600$ kg/m³ fits the measured $\xi = 0.3308$ (and $a = 6387$ km), while $\rho_{03} = 3790$ kg/m³ fits the mean radius $a = 6374$ km (and $\xi = 0.3319$). A compromise value of $\rho_{03} = 3700$ kg/m³ is used here—which gives a fit within acceptable limits, for example with $a = 6378$ km, the equatorial radius, we find $\xi = 0.3312$.

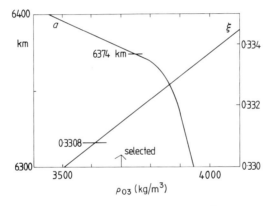

Fig. 7.6. Earth: structural model, choice of ρ_{03} (kg/m³). Model values of radius a (km) and moment of inertia factor ξ.

B. Mars Identification

The data we have are $a = 3400$ km and $\xi = 0.3752$. With the pool model initial value of $f = 80$ we expect the compressibility parameters for the rock substance, especially for the core material, to be different from that of the Earth. The best we can do in selecting these parameters is to assume that the mantle layers will have the same values as those of the Earth, using however the ρ_0 values from the pool model, but choosing the dominant parameter C_4, of the core liquid to provide a fit to (a, ξ). As indicated on Fig. 7.7 there is distinct fit with $C_4 = 79$—indeed the fit is sufficiently distinct to give considerable confidence in this value. (Note also that the Earth value

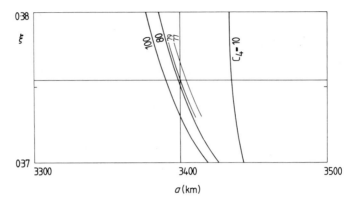

Fig. 7.7. Mars: structural model, choice of C_4. Moment of inertia factor ξ as a function of radius a (km) for various values of C_4.

for C_4, 619, would not allow a fit at all.) The model gives $m_{(core)} = 0.52$ with core radius ratio = 0.77, central pressure = 0.30 Mbar, and central density = 4760 kg/m^3 and pressure at the mantle–core boundary = 0.1 Mbar such that only a layer B is present. The core is surprisingly(?) large.

C. Mercury Identification

We do not have enough information to calibrate a model. All we have, for the given mass, is $a = 2440$ km and the suggested pool value of initial $f = 700$. Clutching at straws, we can however make an attempt by choosing the mantle parameters as for the Earth, but with Mercury pool model values for ρ_0, and guessing a value of C_4. Our guess is guided by the correlation $C_4(f)$ of Fig. 7.5, which for $f = 700$ gives $C_4 = 5000$. This value leads to central densities which seem impractically high. The variation of $m_{(core)}, \xi$ with C_4 is shown in Fig. 7.8. The $I(\bar{\rho})$ correlation suggests $\xi = 0.335$ and hence $C_4 = 1750$. It would be a complete fluke if the $I(\bar{\rho})$ correlation were precise so I choose the round figure $C_4 = 2000$ which gives a not unreasonable set of values for the densities and for which $m_{(core)} = 0.6$. The corresponding core radius fraction = 0.74, central pressure = 0.44 Mbar and central density 9900 kg/m^3, with $\xi = 0.332$. (Note that no model is possible for $C_4 \leq 500$, not too different from the Earth value.)

D. Venus Identification

Here the situation is similar to that of Mercury—not enough information, merely for the given mass the radius $a = 6050$ km. The pool value of initial

V. IDENTIFICATION

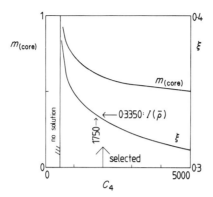

Fig. 7.8. Mercury: structural model, choice of C_4. Core mass fraction $m_{(core)}$ and moment of inertia ratio as a function of C_4.

$f = 250$ gives from the $C_4(f)$ correlation a value $C_4 = 315$. This gives a model fitting the measured radius with $m_{(core)} = 0.42$, core radius ratio $= 0.65$, central pressure $= 2.4$ Mbar and central density $= 9490$ kg/m^3 with $\xi = 0.3395$, and of course (since we have arranged it that way) a full sequence of mantle layers B, C and D.

E. Moon Identification

At first sight making a model of the Moon structure would seem to be straightforward given both $a = 1740$ km and $\xi = 0.392 \pm 0.003$. This is not so. The value of ξ indicates that the role of compressibility in the Moon is very small. The pool model of the Moon suggests also that the zero pressure densities of mantle and core material are not very different so that the body is nearly homogeneous. As a consequence, model calibration is very "loose", namely fits can be obtained over a very wide parameter range with little to suggest that one is much better than another.

The pool model suggests $f = 50$, which with the $C_4(f)$ correlation requires $C_4 = 4$. This value does lead to very small compressibility effects—indeed all values $C_4 \lesssim 10^2$ produce closely similar results. I have therefore chosen the round figure of $C_4 = 10$ for which $m_{(core)} = 0.2$, core radius fraction is 0·565 (the mantle–core boundary being at 760 km depth), central pressure is 0.05 Mbar, mantle density is 3205–3293 kg/m^3 and core density is 3735–3740 kg/m^3, fitting $a = 1740$ km and $\xi = 0.392$.

There is some evidence about the Moon's interior from the work on lunar seismology (see, for example, the discussion in Cook, 1980, pp. 142–153). All the evidence points to a major discontinuity at a depth of 800 km. Whether or not this is the mantle–core boundary as indicated here is an open question. A further possible discontinuity, perhaps owing to an inner core, has been identified at radius 170–360 km.

An insight into the insensitivity of the model calibration can be obtained from what I have called the "egg" model (originally due to E. Wiechart, 1897, and others—see Bullen, 1975, pp. 72–78), namely a uniform mantle of density ρ_m and core of density ρ_c for which

$$\bar{\rho} = \rho_m + \Delta\rho\, r'^3 \qquad \xi'\bar{\rho} = \rho_m + \Delta\rho\, r'^5$$

where $\Delta\rho = \rho_c - \rho_m$, $r' = r/a$ and $\xi' = 5\xi/2 \approx 0.980 \pm 0.008$.

The difficulties arise since ξ' is very close to unity. Given $\xi = 0.392$ and $\bar{\rho} = 3340\,\text{kg/m}^3$ there is an upper limit to ρ_m of $3273\,\text{kg/m}^3$. If we select, for illustration, the value of ρ_m such that $\xi \equiv 0.392$ and $m_{(core)} = 0.200$, as obtained by the full model described above, then $\rho_m = 3242.6\,\text{kg/m}^3$—a reasonable mean value with the full model giving ρ_m of 3205–$3293\,\text{kg/m}^3$. Then for the estimated range of $\xi = 0.392 \pm 0.003$ we obtain core densities $10{,}403$–$3468\,\text{kg/m}^3$ and core mass ratio 0.042–0.448, a wide range! Taking the narrower range $\xi = 0.392 \pm 0.001$ we obtain core densities 4135–$3628\,\text{kg/m}^3$ and core mass ratio 0.135–0.275. With the central value $\xi = 0.392$ the core density is $3800\,\text{kg/m}^3$, a little higher than the full model values of 3735–$3741\,\text{kg/m}^3$.

For a core mean density of $3738\,\text{kg/m}^3$, as in the full model, we require $\xi = 0.3924$. Thus very small changes in the parameters lead to large changes in the model and we cannot therefore be at all confident about particular results.

VI. COMPARISON OF THE TERRESTRIAL PLANETS NOW

Before proceeding it is of interest to compare the present global structure of the terrestrial planets.

The data for the planets now as discussed above is shown in Fig. 7.9, all drawn to the same scale. The Moon and Mars show negligible effects of self-compression whereas those effects are strong for Mercury, Venus and the Earth. The Earth and Venus are seen as the terrestrial planet "twins", and Mercury as being rather different from all the rest as a heavy little planet. All have liquid cores. The detailed mantle structure is seen to be a rather minor feature of the gross structure.

VII. PHYSICAL STRUCTURE SEQUENCE

We now have all the information necessary to evaluate the structure throughout its development as the core mass ratio $m_{(core)}$ *decreases* from

VII. PHYSICAL STRUCTURE SEQUENCE

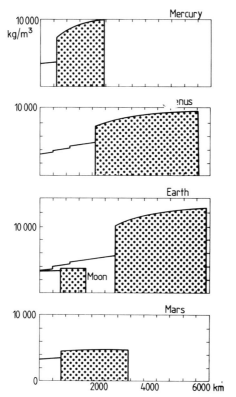

Fig. 7.9. Comparison of the density structure of the present terrestrial planets. All to the same scale, core shaded.

unity to zero. The consequent variation of (a, ξ) for the terrestrial planets is shown in Fig. 7.10. Notice:

1. The radius increases—the liquid is denser than the mantle extract. The change is substantial, being 5420–6805 km for the Earth.
2. The moment of inertia ratio is always substantially less than 0.4, the value for a homogeneous body—except for the Moon, compressibility effects are strong.
3. There is a strong variation of ξ with a minimum, near $m_{(core)} \approx 0.6$, for the Earth. This minimum is not, however, deep enough to make the moment of inertia anything other than a monotonic function increasing throughout the development as shown in Fig. 7.11. For the Earth, the moment of inertia increases from $I/I_* = 0.6$ to 1.0 (where I_* is the value at $m_{(core)} = 0$). This will have a powerful influence on the period of rotation, leading to a slowing of the rate

148   7. PHYSICAL DEVELOPMENT

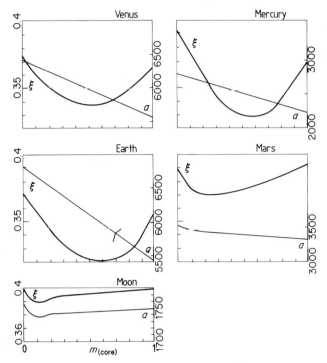

Fig. 7.10. Development of the size and moment of inertia of the terrestrial planets. Variation of the radius a (km) and moment of inertia factor ξ as a function of core mass ratio.

of rotation (longer days and fewer days/year now than in the distant past), but may be swamped by the other powerful effect of tidal interaction. Indeed for Mercury, Venus and the Moon tidal effects have locked their rotation periods to the period of revolution about the central body. (The relatively small change in moment of inertia for Mars is noted in passing.) These very interesting matters are not considered in this book.

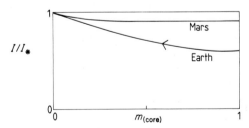

Fig. 7.11. Variation of moment of inertia of Earth and Mars as a function of core mass ratio.

VII. PHYSICAL STRUCTURE SEQUENCE

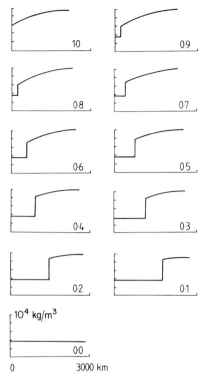

Fig. 7.12. Mercury: density development. Density distribution as a function of core mass ratio.

The radial structure of the Earth (and the other terrestrial planets) is shown in Figs 7.12–7.15 for values of the core mass 0(0.1)1. Let me make a few comments, referring to the Earth model.

1. The body starts its geological life fully molten, $m_{(core)} = 1$. The body is at its most compact and the effect of compressibility is extreme.
2. $m_{(core)} = 0.9$. Mantle layer B is already fully formed and layer C is emerging. The mantle–core density jump is relatively small.
3. $m_{(core)} = 0.8$. Mantle layer C is fully formed and layer D is emerging.
4. The subsequent development has a growing mantle layer D while the core diminishes with slowly falling densities, compressibility effects dominating over the increasing zero pressure density of the core. The mantle–core density contrast goes through a maximum and then slowly diminishes although never to zero.
5. Finally, in a hypothetical state never to be reached, the entire body is solid, layer D reaching to the centre.

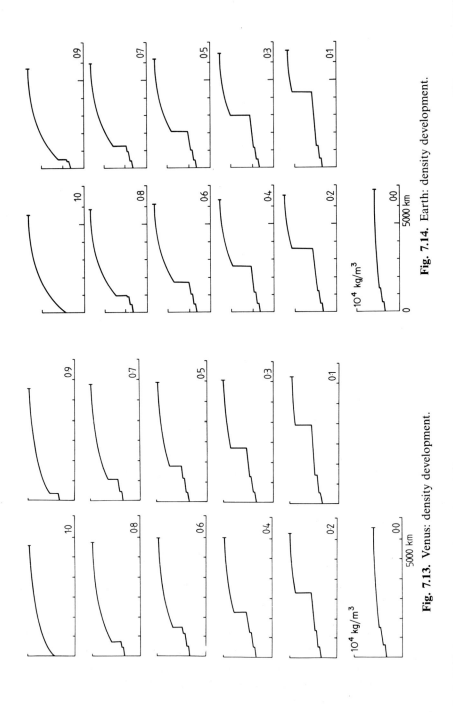

Fig. 7.14. Earth: density development.

Fig. 7.13. Venus: density development.

VIII. THE JOVIAN MOONS

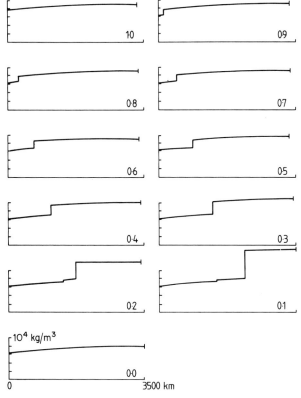

Fig. 7.15. Mars: density development.

A similar pattern is found for the other terrestrial planets except that layers C and D may not form if pressures are not sufficiently high.

VIII. THE JOVIAN MOONS

Our main interest here is the development of the terrestrial planets, the solid "satellites" of the Sun, but it would be remiss not to make some comment in a work of this kind on the other major solid bodies, the satellites of the Jovian planets. So we pause for a moment to consider the structure of these bodies.

The location of the moons of the Jovian planets on the $\bar{\rho}(P_s)$ diagram indicates that they are quite different in composition from the terrestrial planets. With the low (surface) temperatures, about 100 K, of these bodies various frozen "volatile" substances could be considered as the principal constituents. Ice is the obvious candidate; after H and He the next most abundant atom is O, which is used in the brief discussion here, but we need to remember that other low-density materials are available candidates.

A. The Properties of Ice

The phase diagram for ice is shown in Fig. 7.16. (A full description is given by Hobbs, 1974, pp. 60–67, 256–264, 346–351). Ten distinct solid phases have so far been identified. Our interest is in the icy moons of Jupiter and Saturn—the data for the moons of Uranus and Neptune is inadequate—with near-surface temperatures of about 120 and 90 K and pressures up to 10^2 kbar. The phase diagram is reasonably established to about 30 kbar

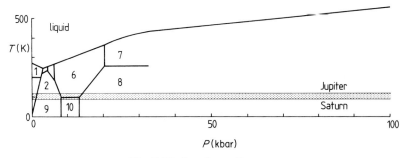

Fig. 7.16. Ice phase diagram.

sufficient to cover all the smaller moons but the ice data is largely unknown for the region of the larger moons, Titan and the Gallileans. Noting that the volumetric change for Ice 6–Ice 8 is very small, and shutting our eyes to Ice 10, the phases of interest are Ice 1, 2 and 6/8 with densities, near 100 K and 1 bar, of 932, 1170 and 1500 kg/m^3 and transition pressures of 0.9 and 9 kbar. If, however, the interior temperatures are considerably greater than 100 K, then the transition pressure could be 2.1 and 6.3 kbar. Hence, for example, if the moons are warm all the small moons of Saturn would have ice largely as Ice 1.

The compressibility χ is about 10^{-5} bar^{-1}. We note that this is an order of magnitude larger than that for rock substance with $\chi \approx (0.4–1) \times 10^{-6}$ bar^{-1}, so that the effects of compressibility will be important for relatively small icy bodies. Unfortunately the data on compressibility are rather uncertain and limited.

The equation of state is very uncertain, especially in the region of interest of 10^2 kbar. I am obliged to do some guessing. In the form $\rho^n = \rho_0^n + nCP$ we need values for n and C. We have $n = 5/3$ for a gas and $n \approx 3$ for rock

VIII. THE JOVIAN MOONS 153

substance. Ice as manifest by its many phases has a relatively "loose" structure so I pick $n = 2$ between that of a gas and a silicate solid. Then the preferred value $\chi_0 = 1.1 \times 10^{-5}\,\text{bar}^{-1}$ gives for $C \equiv \chi_0 \rho_0^n$: Ice 2, 1.5×10^{-4}; Ice 6/8, $2.5 \times 10^{-4}\,(\text{kg/m}^3)^2/\text{Pa}$. (I will ignore the compressibility of Ice 1 since the pressure range is very small.) As we shall see, the models below suggest possible values $C_2 = 4.2 \times 10^{-4}$ and $C_8 = 7.0 \times 10^{-4}\,(\text{kg/m}^3)^2/\text{Pa}$, nearly three times the values from laboratory data. These values are probably extreme; they require, for example, at 50 kbar $\rho = 3040\,\text{kg/m}^3$ rather than the estimated laboratory value $\rho = 2180\,\text{kg/m}^3$. If nothing else this strongly suggests the need for experimental studies of the equation of state of polycrystalline ice near 10^2 kbar. Furthermore, there may be other ice phases at these high pressures.

B. The Small Ice–Rock Moons

This group of small bodies listed in Table 7.4, all satellites of Saturn, with scale pressures of 0.16–2.42 kbar, even with the high compressibility of ice, have negligible compression. Furthermore, they therefore must be entirely or largely of Ice 1 and stony material. Thus using the "egg" model so that the mean density $\rho = \rho_m + (\rho_c - \rho_m)r'^3$ and central mass fraction $s = r'^3 \rho_c/\rho$, where $\rho_m = 932\,\text{kg/m}^3$ is the density of Ice 1, and the density of the stony material here is taken as $\rho_c = 3300\,\text{kg/m}^3$, we can estimate s.

Values of s range from 0.14 to 0.48 with a mass average of $s = 0.34$. There is no obvious relationship of s with orbital distance from the planet, suggesting a high "turbulence" level in the envelope of Saturn during their formation. Considering the 50:1 mass ratio these bodies are all very similar in composition.

Table 7.4. Composition of the small ice–rock moons

	M (10^{21} kg)	a (km)	$\bar{\rho}$ (kg/m³)	P_s (kbar)	m	r
Mimas	0.0376	195	1210	0.16	0.32	0.49
Enceladus	0.074	250	1130	0.22	0.25	0.44
Tethys	0.626	525	1030	0.82	0.14	0.35
Dione	1.05	560	1430	1.78	0.48	0.59
Rhea	2.28	765	1220	2.42	0.33	0.49
Titan[a]						
Hyperion	0.0138	140	1200	0.31	0.31	0.48
Iapetus	1.93	720	1230	2.21	0.34	0.50

[a] Titan data in Table 7.5.
m, r for $\rho_m = 932\,\text{kg/m}^3$ (Ice 1), $\rho_c = 3300\,\text{kg/m}^3$, incompressible material, egg model; bodies in order of orbital distance from the central body, Saturn.

Pluto. The measured data for this enigmatic body is unreliable with mass about 10^{22} kg and radius perhaps 1200–1900 km. It is little better than a guess that it is an ice-stone similar to the smaller moons of Saturn.

Asteroids. Mass and radius data, and that not at all reliable, is available only for three bodies, Ceres, Vesta and Pallas. On the $\bar{\rho}(P_s)$ diagram these lie between the stones and the ice-stones. From measured densities of meteorites we would expect that many of the asteroids would lie to the left of the diagram near $\bar{\rho} = 3000$ kg/m^3.

The solar mix. Is $s = 0.34$ the sort of value we might expect? If we take the solar abundance data we can make various estimates.

1. After the allocation of O to the "metal" atoms, ignoring C and N allocate the remaining O to H_2O. If there is no segregation this gives the maximum possible H_2O mass fraction of 0.41 and all the remainder is stony.
2. If instead all the excess O is allocated first to CO_2 and the remainder to H_2O we find H_2O fraction $= 0.17$, CO_2 fraction $= 0.58$ and stony fraction $= 0.25$. This is fine for the stony fraction but requires a dominantly CO_2 body—and although it is an interesting possibility that the small moons of Saturn are largely dirty frozen soda-pop, at the moment there is not a bubble of evidence.

The conclusion is inescapable, if not obvious, that substantial segregation has occurred. Clearly this is so when we consider the high densities of the innermost Gallileans, Io and Europa.

C. The Large Ice–Rock Moons Callisto, Ganymede and Titan

Again we have run out of information. Let us guess in the light of their apparent correlation in the $\bar{\rho}(P_s)$ diagram that the composition of the three large moons Callisto, Ganymede and Titan is similar to the small moons of Saturn with a mean stony fraction of 0.34, and adjust the compressibility factors C_2 and C_8 for ice so that the model radius fits the body radius. (As discussed above I take the compressibility to be the same for both phases so that $C_2 \equiv 0.6 C_8$.) Further assume that the stony material is the same as Earth layer B. We find $C_2 = 4.2 \times 10^{-4}$ and $C_8 = 7.0 \times 10^{-4}$ (kg/m^3)^2/Pa. The model structures are shown in Fig. 7.17 and properties listed in Table 7.5. These are three remarkably similar bodies with a mantle of Ice 1, 2 and 8 of depth about 1000 km. The effect of the extreme(?) choice of ice compressibility is apparent in the mantle density gradients compared to that of the stony central region where, on the scale of the diagram, the density gradient is barely apparent.

VIII. THE JOVIAN MOONS

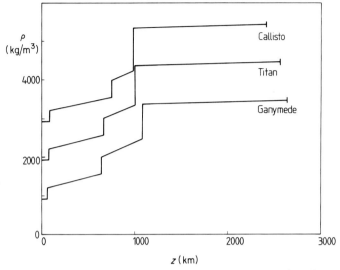

Fig. 7.17. Density structure for Callisto, Titan and Ganymede. The curves for Titan and Ganymede are displaced by 1000 and 2000 units, respectively.

D. The Large Stony Moons Europa and Io

From the $\bar{\rho}(P_s)$ diagram the moons Europa and Io are seen as being quantitatively different from the other moons of Jupiter and Saturn, indeed they lie near the Moon. Clearly they are composed predominantly of rock substance. Io is too dense for this to be entirely our favourite Earth layer B. There are now so many possibilities that models of these two bodies are

Table 7.5. Composition of the large ice–rock moons

	M (10^{21} kg)	a (km)	$\bar{\rho}$ (kg/m³)	P_s (kbar)	m	r
Titan	135.9	2560	1930	68.5	0.38	0.61
Io	89.2	1816	3550	116.4	0.95	0.95
Europa	48.7	1563	3050	63.4	0.88	0.88
Ganymede	149	2638	1940	73.0	0.34	0.59
Callisto	106.4	2410	1820	53.4	0.37	0.59
Moon	73.5	1738	3340	94.3	1.00	1.00

1. Compressible material model, details in text.
2. Moon included for comparison.
3. Io, Europa estimates of m, r for $\rho_{04} = 4000$ kg/m³, Titan for $\rho_{04} = 3300$ kg/m³.
4. Model values of I/Ma^2 are respectively 0.33, 0.37, 0.34, 0.33 and 0.33 but are strongly dependent on the assumed structure.

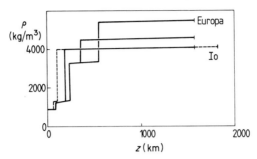

Fig. 7.18. Density structure for Io and Europa. Various possible structures are shown.

quite arbitrary. Purely for the sake of illustration and comparison with the other large moons I show possible model structures in Fig. 7.18 for an ice crust, together with a stony central body. (1) For stony material of Earth type D for both moons the ice crust is composed of Ice 1 and 2 only, since pressures in the crust do not reach the Ice 6/8 transition and the stony fraction is 0.88 and 0.95, respectively. The curve for stony fraction = 0.9 is shown on the $\bar{\rho}(P_s)$ diagram for this class of model. (2) For a stony interior of Earth types B and D a set of possible Europa models is also shown for a given stony fraction of 0.88. All these models suggest an ice mantle of thickness 100–200 km. (This being so the role of compressibility in these two bodies is negligibly small in contrast to Callisto, Ganymede and Titan.)

Finally, we are just left with intriguing speculations. Do these large moons have liquid cores? Is the core material "wet"? Does the mantle in the stony central part have a mineral assemblage like the high pressure low temperature metamorphic assemblages found on Earth? What other frozen volatiles do they preserve in their icy mantles as a record of their origin? What is the thermal history of these bodies?

CHAPTER 8
Thermal History

I. INTRODUCTION

The global structures of the terrestrial planets develop through time. Representing these is a formidable task. As has already been discussed, there are three separate, though related, aspects: chemical, physical and thermodynamic development. For the Jovian planets, at the level of detail in this book, matters are straightforward: there is no chemical development; the structural form remains the same; and the thermodynamic development is simply a consequence of the contraction of the body with a length scale proportional to $t^{-1/3}$. For the terrestrial planets all three aspects have their own strong characteristic features.

Chapters 6 and 7 have presented a sketch of the development of a terrestrial planet as a chemical and physical structural sequence, an ordered arrangement of structures, but there has been minimal explicit reference to the relation of the sequence to time. As well as rearranging the matter distribution the system is rearranging its energy distribution and can only progress from one state to the next insofar as the energy can be transported through the system. The amounts and rate at which energy can be transported determine the time scale of the rearrangement. So let us now consider the temporal development of the structure of the terrestrial planets.

A thermal history focuses attention on the evolution of the gross thermodynamic structure of a planetary body. The model can be calibrated, at least in part, against measurable features dependent on the thermodynamic state of the planet.

II. MODEL SYSTEM

Consider a model terrestrial planet losing heat from its interior through the surface with an interior hotter than the surface and the surface temperature maintained by external processes; see Fig. 8.1. The body is initially fully molten and develops a mantle at a rate determined by the available energy resources and by the heat transfer rate through the mantle. The heat transfer mechanism is assumed to be convection in the "solid" mantle. Two structural features are of major interest: the growth of the mantle characterized by the diminishing radius of the liquid core; and, high in the mantle, a zone of partial melting which diminishes in vertical extent and ceases to exist after a finite time.

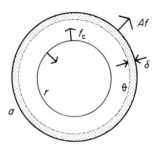

Fig. 8.1. Schema for thermal development.

Conservation of energy of the whole body requires

$$d(\rho V E)/dt = -Af$$

where A and V are the surface area and volume of the body, E is the mean total specific energy, ρ is the mean density and f is the net outward surface heat flux.

Conservation of energy at the mantle–core interface requires

$$H\frac{dr}{dt} = -f_c$$

where H is the energy released per unit mass in freezing at the core–mantle interface and f_c is the net heat flux out of the core.

The quantities E, H, f and f_c are determined by the properties of the matter, and the structure and temperature distribution of the body. In this lumped parameter model they are functions of the core radius r and the global representative temperature θ. Thus given initial values of θ and r the above relations give $\theta(t), r(t)$.

III. ENERGY CONTRIBUTIONS

(Note that in this model the radius of the planet is taken as constant. The effect of changing radius, as described in Chapter 7, is however a contributor to the gravitational energy released and this is taken into account.)

Initial and Boundary Conditions

At the end of the proto-planetary stage a planet is an entirely liquid body. Ultimately the first permanent crust begins to form. This moment, taken as 5 Ga ago, is time zero in the thermal history models presented here. Thus at $t = 0$, the temperature $\theta = \theta_0$ where θ_0 is presumed known; the core radius $r = a$, the radius of the planet; and $E = E_0$.

The surface temperature θ_s is presumed to be constant in time (determined by the solar energy input and the response of the atmosphere–hydrosphere). The values used are: Mercury, 445 K; Venus, 750 K; Earth and the Moon, 300 K; and Mars, 225 K. Possible variation of order 10^2 K in these values have minor effects on the behaviour of the model.

Although the role of the surface temperature is minor (relative to the excess temperature of the interior), because of its influence on the temperatures of the outer parts of a planet there are noticeable effects. This is particularly apparent when we compare Venus and Earth.

III. ENERGY CONTRIBUTIONS

The major energy resources are: the internal thermal energy, with the extra contribution from the core arising from its latent heat and excess temperature treated separately; and the gravitational energy. The minor contribution of radioactive material is presumed to be confined to a thin zone near the surface and not to affect the internal budget.

The form of the energy contributions is stated baldly here and is related to the thermal structure in the following sections.

A. Contribution from Internal Thermal Energy

The thermal energy is a major resource. It is necessary to relate the specific thermal energy to the temperature. In a simple model such as this, in which we ignore the details by in effect integrating over space to obtain a lumped parameter model, the choice of a representative temperature is not straightforward. I choose the "global representative temperature" θ such that the total internal thermal energy for a planet of mass M and nominal specific heat $c = 1$ kJ/(kg K) is $Mc\theta$. Furthermore, this representative temperature is for a body taken to be everywhere below the solidus.

B. Contribution from the Thermal Energy of the Core

The core material is different from that of the solid mantle; except at the core–mantle interface it will have a higher temperature than any solid extract, and in the liquid state carries the latent heat of melting. This extra thermal energy, over that of a mantle solid, is modelled as the energy per unit mass H.

An order of magnitude estimate of H can be made:

$$H = L + (c_c - c_m)\theta_m + c_c \Delta\theta$$

where θ_m is the temperature at the core–mantle interface, $\Delta\theta$ is the temperature difference across a transition zone between the fully solid mantle and the fully liquid core, and L is the latent heat of freezing. For example, with $L = 0.5$ MJ/kg, $c_c = 1$ kJ/(kg K), $c_m = 1$ kJ/(kg K) and $\Delta\theta = 500$ K we obtain $H = 1$ MJ/kg. The model values of H obtained below increase with planet size and core–mantle density contrast largely through the effect of $\Delta\theta$.

The quantity H is a phenomenological constant for each planet. The corresponding contribution to the global specific energy is taken to be $H(r/a)^3$—crudely proportional to the mass of the core.

C. Gravitational Energy Available from Structural Change

As a planetary body rearranges its mass distribution—with a growing mantle and increasing radius—its gravitational energy becomes smaller, the energy change providing a source of thermal energy. The gravitational work function has the form $W = \tfrac{3}{5}\zeta GM^2/a$, where ζ is a dimensionless quantity close to unity dependent on the mass distribution. It is convenient to use as a

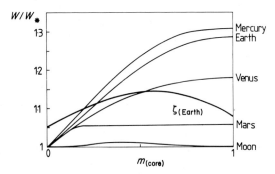

Fig. 8.2. Gravitational work function ratio W/W_* as a function of core mass $m_{(\text{core})}$. The dimensionless ratio ζ for the Earth is also shown.

III. ENERGY CONTRIBUTIONS

reference value the work function W_* corresponding to that of a completely frozen body, $W_* = \frac{3}{5}\zeta_* GM^2/a_*$.

The quantity W/W_* for the terrestrial planets is shown in Fig. 8.2 and related data in Table 8.1. There are a number of features to notice: the change in W for Mars and the Moon is small and negligible for the past interval of geological time and will be ignored for these two bodies; and the change in W is substantial for Mercury, Venus and Earth.

Table 8.1. Gravitational work function

	a_*	ζ_*	(W/W_*)
Mercury	2800	1.0413	1.305
Venus	6443	1.0508	1.18
Earth	6805	1.0527	1.286
Moon[a]	1740	1.00	1.00
Mars	3524	1.0310	1.055

[a] Moon values for reference radius $a = 1740$ km, present value, with change in W/W_* over range less than 0.6%.

The energy per unit mass $w \equiv W/M$, e.g. for the Earth is about 40 MJ/kg. This is very large compared to the other energy sources. The available fraction is 30% so that about 12 MJ/kg is available to run the global system.

D. Energy Total

The total specific energy is

$$E = c\theta + H(r/a)^3 + w$$

comprising the contributions of thermal, latent and gravitational energy. The various terms have typical magnitudes in units of MJ/kg, respectively 3, 5 and 10 for Earth and Venus. Of the three contributors only the thermal energy is a strong contributor for all the planets throughout geological time, the role of latent heat being important early in geological time, and the role of gravitational energy being important only for the large terrestrial planets Venus and Earth.

In the numerical models, from known values of θ and r, new values are obtained at a later time by: evaluating f; finding the new E from the whole body energy equation, so that $\theta = (E - H(r/a)^3 - w)/c$; obtaining the new core radius as described in the following section; and finally evaluating any quantities of interest.

IV. THE CORE RADIUS

The rate of development of the model system is determined by the two fluxes f and f_c. The flux f is determined by heat transfer in the mantle and can be obtained readily, in principle, from measured properties of the upper mantle (see below). The flux f_c could be determined similarly but there is no information on the relevant parameters. Let us therefore make the strong assumption that the net loss from the body as a whole and the core arises crudely in proportion to their masses. Then the core heat flux

$$f_c = (r/a)f$$

Combining this with the above conservation relations, provided H is a constant, gives

$$\log(r/a) = -\frac{1}{3H}(E_0 - E)$$

Thus, in a numerical representation of the model there is a single relation to integrate forward in time to obtain a new value of E, so that, given the relation $E(0)$, together with that for $\log(r/a)$, new values of θ and r are obtained.

It is of interest to note that this result provides a method of obtaining the mean surface heat flux $\bar{f}$ over geological time. The net loss of energy ΔE (per unit mass) from the interior in a time interval Δt requires

$$\bar{f} = \tfrac{1}{3}\rho a\, \Delta E/\Delta t = -aH \log(r/a)/\Delta t$$

With $t = 5$ Ga, using the values of r estimated in Chapter 7, and using the values of H found below and given in Table 8.3, mean fluxes for Mercury, Venus, Earth, the Moon and Mars, respectively, in mW/m^2 are 91, 447, 860, 21 and 67. (We regard each planet as a large calorimeter with the energy content read from the core radius.)

Furthermore, as we shall see, the initial rate of development of the mantle is rapid so that the core reaches close to its present size early in its geological life. If the initial surface heat flux is f_0 we have the time scale of the core radius $\tau_c = \rho a H/f_0$, so that initially $r = a \exp(-t/\tau_c)$. For the selected models described below, τ_c in Ga is: Mercury, 1.7; Venus, 1.0; Earth, 1.0; the Moon, 1.7; and Mars, 2.6.

V. TEMPERATURE DISTRIBUTION

Consider the mean temperature profile at a particular instant as sketched in Fig. 8.3. The profile of interest is bounded by two lines: (1) the liquidus of

V. TEMPERATURE DISTRIBUTION

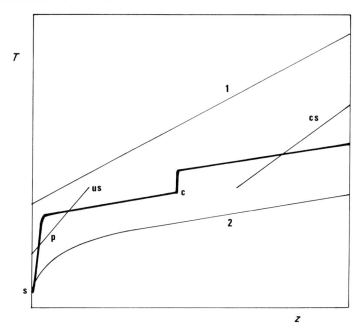

Fig. 8.3. Schematic temperature profile.

the original undifferentiated material, of a fully liquid body, at geological time $t = 0$; and (2) a sub-solidus of a fully solid body, at a hypothetical time, not reached by the terrestrial planets. At the surface (s), for all $t \geqslant 0$, the temperature is determined by the solar flux, the planet's orbital distance and the state of the atmosphere—and is taken here as independent of time. The profile has two distinct parts, those of the solid mantle and the liquid core, with a common temperature at the mantle–core interface (c). The form of the profile will be determined by the properties of the material and the intensity of the heat transfer processes. As time elapses and energy is lost the temperatures fall.

The following points are relevant to the discussion below:

1. The form of curve (1) will be that of the melting point for depth (MPD) relation—a column everywhere at its melting point, with gradients of order 1 K/km.
2. The form of curve (2) will be determined by the entire thermal history up to the moment all the material is frozen.
3. The mantle temperature profile will be determined by "solid" convection in the mantle (details below).

4. Note that there will be an initial stage during which there will be a region in which mantle temperatures exceed the solidus temperatures (us) so that a zone of partial melting (p) will exist.
5. The core will in general be hotter than the mantle. The bulk of this excess temperature presumably will be taken up in a thin boundary layer at the top of the core—a mush zone of thickness of order 10 km for the present Earth with $f_c \approx 8\,\text{mW/m}^2$ (assuming the core radius model of Section IV).
6. The inner core will be at temperatures below the core material-extract-solidus (cs). This zone is ignored here.
7. Otherwise both the deep mantle and core temperatures, on the assumption they are well stirred by convection, will be close to adiabatic (with small gradients of order 0.1 K/km).

A. The Initial Global Mean Temperature θ_0

The increment to the initial mean global temperature which arises from the increase of melting point with pressure can be estimated roughly by considering the liquid sphere to be homogeneous—namely, ignoring the role of compressibility. Thus the pressure $P = P_c(1 - r'^2)$, where $P_c = \frac{2}{3}\pi G \rho^2 a^2$ and $r' = r/a$; the melting point temperature $T = T_0 + \beta P$; and the mass mean melting point temperature $\bar{T} = \int T\,dm / \int dm = T_0 + \frac{2}{5}\beta P_c$, an increment of $\frac{2}{5}\beta P_c$. For example, with $\beta = 10^{-3}$ K/bar for Mercury, Venus, Earth, the Moon and Mars the increment (in K) is 100, 550, 700, 20 and 100, respectively. There is insufficient information to specify these quantities reliably. In the illustrative model (of Section VIII) I have therefore chosen nominal values of $T_0 = 2400$ K with increments of 100 K for Mercury and Mars, and 600 K for Venus and Earth, to give θ_0: Mercury, 2500 K; Venus, 3000 K; Earth, 3000 K; the Moon, 2400 K; and Mars, 2500 K. (Numerical experiments with the model show that increments at least as great as three times the illustrative values would be possible without leading to difficulties of the model in being compatible with the known data.)

B. The Upper Mantle Temperature Profile

The structure of the upper mantle temperature field is presumed to be similar to that of a layer of fluid losing heat from its upper surface and in a vigorous state of thermal convection (for details see, for example, Elder, 1981). Laboratory measurements of such systems reveal a temperature field continually varying in space and fluctuating in time. The variations and fluctuations are of amplitude about 10% of the total temperature difference across the system.

VI. PARTIAL MELTING

The mean temperature profile has a thin near-surface zone in which the vertical temperature gradient is large, with measured near-surface values on Earth of typically 25 K/km. It is convenient to define a length scale for this zone such that the heat flux $f \equiv K\Delta T/\delta$, where ΔT is the temperature difference across the sublayer: measurement shows that the temperature increase at depth δ is about $0.80\Delta T$, and the full temperature increase ΔT is reached at a depth of about 3δ.

The temperature profile $\vartheta(z)$ used here is obtained near the surface from the form found by measurement in thermal turbulence and at depth from an adiabatic profile tangent to that of the surface zone. Thus:

(1) $0 \leqslant z \leqslant z'$ writing $\phi = \vartheta/\Delta T$ $y = z/\delta$
$$\phi = \phi(y) = 1 - \exp[-(y + 0.6y^2)]$$
(2) $z \geqslant z'$ $\vartheta = \vartheta' \exp(\gamma g z/c)$

where z' and ϑ' are chosen such that the adiabatic temperature gradient $d\vartheta'/dz = \gamma g \vartheta/c \approx \gamma g \Delta T/c$ is equal to that given by ϕ. This is readily achieved by iteration. In general the adiabatic gradient is small (~ 0.1 K/km) so that the fit of these two functions is in the lower part of the sublayer with $z' \gg \delta$ and $\vartheta \approx \Delta T$. (The dimensionless adiabatic gradient $\gamma g \delta/c \sim 0.01$, to be compared with the dimensionless surface gradient of unity.)

C. The Sublayer Base Temperature

These profiles are used in this work only to determine the amount of partial melting. Otherwise outside the thin sublayer the temperatures are adiabatic. But one further item is needed in the model to link the various representations of the temperature field, namely the relation of the temperature ΔI across the sublayer to the global representative temperature θ. Ignoring the thin sublayer, this is the mass average of the adiabatic temperature of mantle material for the whole body:

$$\theta = \theta_s + \frac{3}{a^2} \int_0^a \vartheta r^2 \, dr$$

where ϑ is the adiabatic temperature.

Thus $(\theta - \theta_s)/\Delta T \equiv \xi$ is a function of $\gamma g/c$ which, e.g. for Earth with $g = 10\,\text{m}^2/\text{s}$, $c = 1\,\text{kJ}/(\text{kg K})$ and $\gamma = (1, 2) \times 10^{-5}\,\text{K}^{-1}$, gives $\xi = (1.18, 1.42)$. In this work I have taken $\xi = 1.25$, the same for *all* the terrestrial planets.

VI. PARTIAL MELTING

The major process which occurs throughout geological time is the growth of the mantle as a consequence of the phase changes at the mantle–core boundary. Vigorous processes also occur within the solid mantle and crust as the system continually rearranges itself. Two major internal processes contribute to this rearrangement: convection of the solid mantle and the production of magmas from zones of partial melting within the mantle.

Partial melting does not continue indefinitely. It has already ceased on some of the terrestrial bodies. Data from Earth-based observation but especially from material and data collected from landings on the Moon and Mars provide evidence about their geological history. Volcanism ceased long ago on the Moon and probably also on Mercury and Mars. The cessation of volcanism provides a characteristic time marker with which we can calibrate our ideas about structural development. The size of surface volcanic systems also provides valuable clues.

Consider our model thermal history from which we obtain the temperature profile $\vartheta(z)$ and in which the role of partial melting is ignored. If, however, $\vartheta > T_m$, the local melting temperature, a portion of the material will be molten. Let the volumetric fraction in the molten state be $e = e(z)$. Conservation of energy then requires

$$\rho_* c_* \vartheta = (1-e)\rho_* c_* T_m + e\rho(L + cT_m)$$

where ρ_* and c_* refer to the solid and ρ, c and L to the liquid, and L is the latent heat of melting.

The actual temperature will be T_m, whereas ϑ is the temperature that would have arisen if melting were not permitted. It is convenient to write $\rho = (1-\xi)\rho_*$ and make the approximation $c = c_*$, whence

$$e = (\vartheta - T_m)/\vartheta_*$$

with $\vartheta_* = (1-\xi)L/c_* - \xi T_m$. (For example, with $L = 3$ MJ/kg, $c_* = 1$ kJ/(kg K), $T_m = 10^3$ K and $\xi = 0.1$ the quantity ϑ_*, a property of the rock substance alone, is 2600 K.)

Notice the following consequences:

1. $\vartheta \leq T_m$ requires $e = 0$;
2. for $\vartheta = \vartheta_1$, where $\vartheta_1 = T_m + \vartheta_*$, $e = 1$;
3. for $T_m < \vartheta < \vartheta_1$, $0 < e < 1$;
4. for $\vartheta > \vartheta_1$, we have $e = 1$ and fluid temperatures above that of the liquids.

The MPD is taken as $T_m = \alpha + \beta P$. Since our major interest, for the purpose of model identification, is in whether or not partial melting occurs at all I choose $(\alpha, \beta) = (1400 \text{ K}, 6 \times 10^{-3} \text{ K/bar})$ values appropriate to a typical basaltic magma (other magmas will exist but if the system cannot produce basalt it is volcanically dead—or very close to it).

VI. PARTIAL MELTING

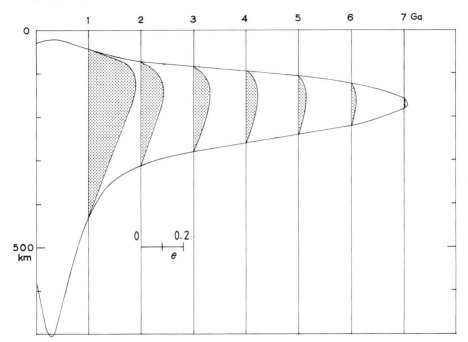

Fig. 8.4. Partial melt development. Profiles of volumetric melt fraction e at time 1(1) 7 Ga for the Earth.

Model results for the Earth are shown in Fig. 8.4. The system starts with a completely molten mantle and ends with a completely solid one. For the Earth, over an interval of about 7 Ga, there is a zone of partial melting of restricted vertical extent located in the upper portion of the mantle, and the extent of this zone diminishes with time.

Role of Temperature Fluctuations

Superimposed on the mean temperature θ are temperature fluctuations which vary in space as well as time and have amplitude θ' which from laboratory measurements of thermal turbulence are of order 0.1θ. The depth to the top of the partially melted zone for the Earth at $t = 5$ Ga with model values is shown in Fig. 8.5, together with the corresponding value of e_{max}, the maximum value of $e(z)$. For $\theta' < -110$ K no melting is possible, otherwise the melting depth falls rapidly near $\theta' = 0$; but for $\theta' > 400$ K the change in melting depth is small. This suggests that, for the present Earth, partial melting of the upper mantle is confined to depths greater than about 70 km, can be as deep as about 175 km, and for expected fluctuations of 200–300 K

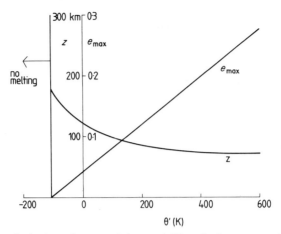

Fig. 8.5. Depth (km) to the top of the partially melted zone as a function of the temperature fluctuation (K). Also shown: the maximum over the distribution with depth of the volumetric partial melt fraction e_{max}.

will be predominantly at depths of about 80–90 km with e in the range 0.12 to 0.16, to be compared with the global value at $\theta' = 0$ of $e = 0.04$. These depth ranges are not too different from those suggested by the study of the petrogenesis of the parent magmas of basalts and of kimberlites.

Thus in general partial melting will occur sporadically and with varying intensity—not necessarily being present everywhere—within a shell of restricted radial extent; a shell whose extent diminishes with time.

VII. HEAT TRANSFER

The net effect of the many distinct heat transfer processes within the interior of the planetary body is to produce an outward surface heat flux. Three processes are envisaged as the main contributors: (1) free convection of magma in the zone of partial melting, aided in the case of the Earth by the circulation of water substance in the near-surface zone; (2) free convection of the "solid" mantle; and (3) transfer by thermal conduction of the heat released by radioactive decay of radiogenic materials presumed to be concentrated in the crust.

A. Magma Convection

As we shall see (Section VIII), "solid" convection alone can provide an adequate transport mechanism for the smaller terrestrial planets. This is not

VII. HEAT TRANSFER

the case for the larger bodies Venus and Earth, for which the initial energy stocks are large owing to the large contribution from gravitational energy. For example, the Earth has an excess of about 10 MJ/kg which cannot be removed by "solid" convection.

As we have seen (Section VI), all bodies pass through a stage in which part or all of the mantle is partially molten. Measurements of modern active hydrothermal systems, in which the working fluid—water substance—circulates freely in the crust, reveal heat transfer rates typically 10^2 to 10^3 those of simple conduction (see, for example, Elder, 1981). By analogy, the free circulation of magma in zones of partial melting could produce similarly enhanced heat transfer rates of typically 10 W/m^2. Rates of this order would be sufficient to transfer the excess available energy.

The role of this magma convection process will not, however, be confined to the larger bodies—it will presumably operate for all the terrestrial planets while a zone of partial melting exists.

An attempt at a detailed study of the net effect of free convection in the partial melt zone would be not only beyond present knowledge but also could not be calibrated. For the moment it is sufficient to note that a key factor in the operation of this type of convection is the permeability of the medium. The permeability is a strongly varying function of the porosity, in this case the volumetric melt fraction e, varying at least as strongly as e^3. Thus early in the geological history when e is largest the process will be vigorous; it will fall rapidly in vigour as e falls.

Thus I am forced to represent the effect of the process empirically by presuming its net effect is to produce an enhanced surface heat flux, $f_p = f_p(t)$. I choose $f_p = f_0 e^{-t/\tau_p}$, where f_0 and τ_p are phenomenological constants.

The net effect of this process, after $t \gg \tau_p$, is the removal of a total amount of internal energy per unit planetary mass of $3f_0\tau_p/\rho a$. In the selected models described below this quantity ranges from 0.7 MJ/kg for the Moon to 10.8 MJ/kg for Earth.

In the numerical models described below I have selected a common value for all the terrestrial planets, $\tau_p = 0.4 \text{ Ga}$ (an interval during which there is in general a substantial initial fall in e_{max}), thereby leaving f_0 as a single free parameter. Note that after a time of about $3\tau_p$, for which $f_p/f_0 \approx 0.05$, the bulk of the effect of this process is past with a net effect thereafter for a particular planet dependent on $f_0\tau_p$.

B. "Solid" Convective Heat Transfer

The identification of the second major heat transfer process can be made by considering the net outward surface heat flux of the present Earth. The mean global value is about $f = 50 \text{ mW/m}^2$. To put this value in perspective it is useful to relate it to the corresponding passive state of a model body in which heat is transferred solely by thermal conduction—with a surface heat flux of scale $K\Delta T/a$. The ratio of the actual to this conductive scale value defines the global Nusselt number $N = fa/K\Delta T$ such that if $N \approx 1$ the dominant process is thermal conduction and if $N \gg 1$ some other active process is operating.

An estimate of the present Earth's Nusselt number can be made using $K = 3$ W/(m K); $\Delta T = 1500$ K (a guess at this stage). This gives $N = 70$. Clearly conduction alone is inadequate and some vigorous heat transfer process is operating.

The key question is then what process or processes produce this high heat transfer rate. Elsewhere (Elder, 1981) I give an estimate of the energy budget of the present Earth's upper mantle (global power in units $10^{12}W$): total recharge from the lower mantle, 57; total discharge to the lower mantle of recirculated and cooled upper mantle, 32; loss through the surface, 25—of which the contribution from mass discharge at the surface, largely as surface volcanism, is 1.7. Thus the loss produced by volcanism is at present about 7% of the total loss. The bulk of the loss arises from the effects of mantle convection.

Laboratory studies of vigorous free-convection in a layer of fluid cooling from above show that the rate of heat transfer is increased, over what it would be without convection, by the ratio N, the Nusselt number, which for materials of large Prandtl number (v/κ) is determined principally by the Rayleigh number, $A = \gamma g \Delta T h^3/\kappa v$, where γ is the coefficient of cubical expansion, g is the acceleration of gravity, ΔT is the temperature drop across the convecting layer, h is the depth of the convecting layer, and κ and v are the thermal diffusivity and kinematic viscosity of the convecting fluid. For $A > 10^6$ we find experimentally that the fluid is turbulent and $N = (A/A_c)^{1/3}$, where $A_c \approx 700$–750, taken here as $729 = 9^3$, is a dimensionless quantity obtained by experiment. In these circumstances the power transferred is independent of the layer depth. Furthermore, the mean temperature profile across the layer is such that the bulk of the material is at a uniform temperature (in the laboratory) of mean temperature excess ΔT, and the mean temperature differs from that at the surface only in a thin layer, the sublayer, of thickness $\delta \ll h$. It is found that $\gamma g \Delta T \delta^3/\kappa v = \text{constant} = A_c$ —i.e. the sublayer is in a marginally stable state with a constant sublayer Rayleigh number of A_c. The heat flux and sublayer thickness are

$$f = K\Delta T/\delta \qquad \delta = A_c^{1/3}(\kappa v/\gamma g \Delta T)^{1/3}$$

Thus the heat flux is determined for a given body solely by the properties κ, v and γ, and by the temperature across the sublayer.

VII. HEAT TRANSFER

A major difficulty arises when we attempt to apply the results from simple laboratory systems to planetary bodies, mainly because of lack of knowledge of the properties of the medium. The material properties κ and γ are reasonably well known, perhaps to within better than a factor of 2, but that is not the case for v. In spite of a vast amount of detailed work on measuring post-glacial uplift and related quantities, estimates of the kinematic viscosity range from 10^{16} to 10^{18} m²/s with typically 3×10^{17} m²/s for models with a uniform mantle, and 10^{16} m²/s for models with a relatively low viscosity "channel" of vertical extent of about 100 km (e.g. Mörner, 1980). There is no obvious method for distinguishing between these values.

The kinematic viscosity presents its greatest problem through its variation with temperature. I have used the form

$$v = v_0 \exp b\left(\frac{1}{\theta} - \frac{1}{\theta_*}\right)$$

After considerable numerical experimentation the quantities v_0, b and θ_* for the selected models have been chosen to give values of v which straddle 10^{17} m²/s and which do not have extreme ranges in a particular model. I choose $v_0 = 10^{17}$ m²/s, $b = 10^4$ K and $\theta_* = 2300$ K. This choice, in the selected models described below, leads to a kinematic viscosity range (the ratio over the interval 0–5 Ga) of 3.2 for Venus to 16.8 for the Moon with an average for all the terrestrial planets of 5.2. (Small values of $b < 10^3$ K produce results little different from those of constant kinematic viscosity, $b = 0$; large values of $b > 10^5$ K require unrealistically small initial values of δ, typically of order 1 km.)

In applying these laboratory studies to a planetary body it is convenient to choose the length scale $h = a$, the radius. Then writing $A = \gamma g a^3 \Delta T/\kappa v$ with initial value $A_0 = \gamma g a^3 \Delta T_0/\kappa v$ we have

$$\delta = \delta_0 (\Delta T/\Delta T_0)^{-1/3} \qquad \delta_0 = a(A_c/A_0)^{1/3}$$

Let us also estimate the corresponding global Rayleigh number A for the present Earth using $\gamma = 10^{-5}$ K^{-1}, $\Delta T = 1500$ K, $K = 3$ W/(m K), $\rho = 5520$ kg/m³ and $c = 1$ kJ/(kg K), so that $\kappa = 5.43 \times 10^{-7}$ m²/s and $v = 3 \times 10^{17}$ m²/s. This gives $A = 2.4 \times 10^8$ with convective $N \approx 70$. This value is sufficiently close to the value estimated from measurement to indicate that solid convection is the major contributor to the Earth's surface heat flux today.

The total flux out of the interior of the body is taken as

$$f_{(\text{interior})} = f_0 e^{-t/\tau_p} + K\Delta T/\delta$$

which is the sum of the contributions from magma convection and solid mantle convection.

A simple and useful picture of the mechanism of the sublayer is as follows (originally proposed by L. Howard, see Elder, 1981, pp. 43–51). Suppose (at time $t = 0$) at a particular moment in the vicinity of a localized patch of the surface that deep, hot fluid has penetrated right up to the surface, so that the temperature locally is everywhere the same as at depth, except at the cold surface, and that the local fluid is stationary. Heat will be lost from the fluid in a zone near the surface; the heat will be transferred solely by thermal conduction; a zone of depth of order $(\kappa t)^{1/2}$ will be affected. For an interval of time $0 \leqslant t \leqslant \tau_s$, say, the cooled layer, although statically unstable owing to its greater density, will be dynamically stable owing to the combined stabilizing role of thermal diffusion and viscosity. If, however, the Rayleigh number for the system as a whole is sufficiently large, instabilities within the cooled layer grow to finite amplitude, until the cooled mass falls into the interior and hot, deep fluid is recirculated back towards the surface. The period of gestation in which the cooled layer grows by conduction is followed by a short interval in which the cooled fluid is ejected out of the cooled region to be replaced by deep hot fluid, thereby more or less restoring the conditions near the surface. As seen in the laboratory this process is observed to occur more or less at random over the surface zone.

In order to quantify the gross effects over geological time of the major mass (and heat) transport process, namely the convective recirculation of mantle material into and out of the upper mantle, consider $n = n(t)$, the number of times the material of the upper mantle has been recycled where

$$dn/dt = 1/\tau_s \qquad \tau_s = \alpha \delta^2 / \kappa$$

In the simple model the constant $\alpha = 4/\pi$. In that model no allowance is made for the variation of viscosity with temperature. Also κ is the thermal diffusivity of upper mantle material rather than the global thermal diffusivity based on the global mean density. Rather, let us choose a value of α such that, for the present Earth, a single recycling occurs in a nominal time of 0.3 Ga—an interval during which substantial crustal rearrangement has occurred. We find $\alpha \approx 5$. (Compared with the simple Howard model this implies a single crustal rearrangement after 2.4, 2.5, 2.3, 3.9 and 3.3 recyclings for the selected models of Section IX—Mercury, Venus, Earth, the Moon and Mars, respectively.)

C. Contribution from Radioactivity

There is strong evidence that radiogenic materials are concentrated in the crust and outer parts of the mantle. Therefore the radiogenic contribution is appropriately modelled by means of a modification to the surface boundary conditions rather than by incorporation in the global energy equation.

Consider the case of radiogenic material in a near-surface zone with

VIII. SIMPLE MODEL

power/unit volume $A(z) = A_0 e^{-z/d}$, say, so that the power/unit area of surface is $\Delta f = A_0 d$. The conduction equation then has the solution for temperature:

$$T = T_0 + f_\infty \frac{z}{K} + \frac{A_0 d^2}{K}(1 - e^{-z/d})$$

which satisfies $T = T_0$ on $z = 0$ and $K\, dT/dz = f_\infty$ as $z \to \infty$. The surface heat flux $f_0 = f_\infty + \Delta f$ and the temperature as $z \to \infty$:

$$T = T_0 + f_\infty \frac{z}{K} + \Delta f \frac{d}{K}$$

Since Δf and $\Delta f\, d/K$ are rather small, a precise knowledge of d is not so important.

Both these results require quasi-steady conditions, which will be the case for $d \ll a$. Thus two effects arise:

1. There is an additional surface heat flux $\Delta f = \frac{1}{3}\rho a P$, where P is the global mean specific power.
2. There is an increment to the interior temperature $\Delta \theta = \Delta f\, d/K$, where d is the mean depth of the radiogenic layer (~ 10 km).

Of the major contributors K, U and Th to internal heating by radioactive decay, K is dominant. Thus for a single component the specific radiogenic power (per unit mass) $P = P_0 e^{-\lambda t}$, where $\lambda \approx 0.54$ Ga^{-1} is the decay constant of K. P_0 is the initial power. (In order to have some idea of orders of magnitude, note that a uniform value of $P_0 = 6 \times 10^{-11}$ W/kg in the Earth, provided the energy could reach the surface, would supply the present heat flux of 50 mW/m^2 after 5 Ga. The proportional fall in output $e^{-\lambda t} = 0.067$.)

In this work I (somewhat arbitrarily) assume that no more than 10% of present-day net power output from the Earth's interior is radiogenic. This requires $P_0 = 6 \times 10^{-12}$ W/kg. I use the same value for all the terrestrial planets. Thus the radiogenic contribution to the surface heat flux is $\frac{1}{3}\rho a P_0 e^{-\lambda t}$. Note: this flux is taken from the superficial crustal layer and does not affect the interior energy budget, and the temperature increment is negligible. Thus the net surface heat flux $f_{(\text{surface})} = f_{(\text{interior})} + \frac{1}{3}\rho a P$.

VIII. SIMPLE MODEL

To demonstrate the behaviour of the model system, consider a simplified version in which:

1. the kinematic viscosity is a constant—it does not vary with temperature;

2. the latent heat contribution to the energy equation is negligible—$H/c\theta_0 \ll 1$;
3. the contribution of gravitational energy is negligible—as is the case for Mercury, the Moon and Mars;
4. the contribution of radiogenic power to the surface flux is negligible—$P_0 = 0$; and
5. magma convection does not occur—$f_0 = 0$.

Thus:

$$E = c\theta$$

and

$$d(\rho c V\theta)\,dt = -Af$$
$$f = K\Delta T/\delta \qquad \Delta T = (\theta - \theta_s)/\xi \qquad \delta = \delta(\Delta T)$$
$$\log(r/a) = -(c/3H)(\theta_0 - \theta)$$

with $\theta = \theta_0$, $r = a$ at $t = 0$. The relations for E and r are now decoupled.

Furthermore, the energy equation involves only the variables θ and t and can be immediately integrated. Thus, if we write $K = \rho c \kappa$, so that κ is a thermal diffusivity based on the thermal conductivity of the upper mantle and the mean density of the whole body, the energy equation becomes

$$\xi d\Delta T/dt = -3\kappa\,\Delta T/a\delta$$

which with the convective relation $\delta(\Delta T)$ integrates to

$$\Delta T = \Delta T_0 \phi^{-3} \qquad \phi = \left[1 + \frac{1}{\xi}\left(\frac{A_0}{A_c}\right)^{1/3}\frac{\kappa t}{a^2}\right]$$

The quantities δ, f, r etc. can then be obtained from the relations above. Note in particular that $\delta = \delta_0 \phi$: the sublayer thickness increases linearly with time.

This simple system has a single (thermal) timescale $\tau_\theta = (A_c/A_0)^{1/3} a^2/\kappa$, a strong function of body radius a, and weakly dependent on θ_0—smaller bodies cool faster. (The temperature would fall to $0.5\theta_0$ in a time about $0.2\tau_\theta$.)

Simple Model Behaviour

The behaviour of the simplified model is illustrated in Fig. 8.6. The kinematic viscosity has been chosen as the independent variable since it is the most uncertain parameter and in the full model the kinematic viscosity sweeps

VIII. SIMPLE MODEL 175

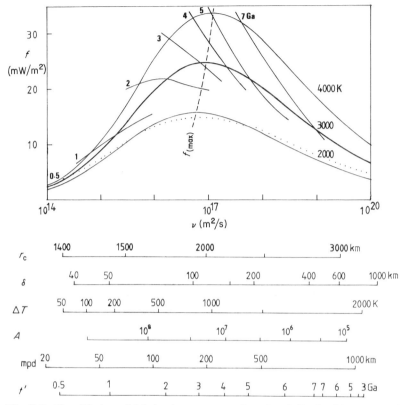

Fig. 8.6. Simplified model behaviour—properties after 5 Ga for Mars. Heat flux f (mW/m^2) as a function of kinematic viscosity v (m^2/s) for initial global representative temperatures 2000, 3000 and 4000 K together with the line of heat flux maxima. The duration of volcanism, in Ga, is also indicated, for $(\alpha, \beta) = (1400 \text{ K}, 2 \times 10^{-3} \text{ K/bar})$; $\theta' = 100$ K. The attached scales refer to the 3000 K data. Note that they apply to $t = 5$ Ga (now)—except for the duration of volcanism scale t'. r_c, core radius; δ, sublayer thickness; ΔT, temperature drop across the sublayer; A, Rayleigh number; mpd, depth to top of zone of partial melting.

over a range of values as the global temperature falls. There are a number of features to notice.

1. There is a broad maximum in f (near 10^{17} m^2/s). With small v convection is strong and the body cools quickly; with large v convection is weak and the body cools slowly. The values of v on the locus of maxima are only weakly dependent on θ_0.
2. The associated quantities θ, δ, r_c and A change monotonically with v, all except A increasing with v.

3. The duration of volcanism has a maximum strongly dependent on θ_0 (of 7.1 Ga, near $v = 10^{19}\,\mathrm{m^2/s}$, with $\theta_0 = 3000\,\mathrm{K}$; 3.3 Ga, near $v = 2 \times 10^{18}\,\mathrm{m^2/s}$, with $\theta_0 = 2500\,\mathrm{K}$; 0.8 Ga, near $v = 10^{17}\,\mathrm{m^2/s}$, with $\theta_0 = 2000\,\mathrm{K}$). Thus, for example, if it were known that volcanism ceased on Mars after 2 Ga, we would require $\theta_0 > 2500\,\mathrm{K}$.
4. Similarly an upper bound to θ_0 can be estimated. For example, for Mars with $\theta_0 = 4000\,\mathrm{K}$ the duration of volcanism is 5 Ga near $v = 10^{17}\,\mathrm{m^2/s}$. But Mars is volcanically dead so that $\theta_0 < 4000\,\mathrm{K}$.

Table 8.2 shows data obtained from the simplified model for a set of model terrestrial planets with the same properties: $v = 10^{17}\,\mathrm{m^2/s}$, $H = 1\,\mathrm{MJ/kg}$ and $\theta_0 = 2750\,\mathrm{K}$ are chosen to give duration of volcanism of 2 Ga on the Moon.

Table 8.2. Simple thermal history model

	r_c (km)	δ (km)	f (mW/m²)	Duration (Ga)	H' (MJ/kg)
Mercury	1560	110	20	3.2	1.50
Venus	4670	80	37	7.0	0.60
Earth	4600	70	51	4.2	0.54
Moon	970	190	9	2.0	1.00
Mars	2130	120	23	2.8	1.80

$\theta_0 = 2750\,\mathrm{K}$, $v = 10^{17}\,\mathrm{m^2/s}$, $H = 1\,\mathrm{MJ/kg}$, $\alpha = 1400\,\mathrm{K}$, $\beta = 6 \times 10^{-3}\,\mathrm{K/bar}$; $\theta' = 100\,\mathrm{K}$. Values at $t = 5\,\mathrm{Ga}$; volcanic duration; and value of H' to give values of r_c equal to those found in Chapter 7.

Although the simplified model applies with least validity to Venus and Earth, they are included for comparison. The behaviour diagram for Mercury, the Moon and Mars is shown in Fig. 8.7. Taken at face value, and if H is the same for all the small terrestrial planets, these model data for the core radii suggest the estimates of core radii made in Chapter 7 for Mars and Mercury to be too high by 15% and 10% of the planet's radius respectively.

If one agreed that Mercury, being a more "basic" body than the Moon and Mars, has a value of (α, β) more appropriate to an ultrabasic material, the duration of volcanism would be shorter. For example, with $\alpha = (1400, 100, 1800)\,\mathrm{K}$ and $\beta = (2, -0.5, 0) \times 10^{-3}\,\mathrm{K/bar}$, the duration is $(3.2, 2.9, 2.5, 2.2, 2.0)\,\mathrm{Ga}$.

The fit of the simple model to Venus and Earth is poor—the core radii, for example, are too large; and volcanism on Earth would have ceased. Several features contribute to this inadequacy, notably the neglected gravitational energy and the role of H in the energy equation. The fit to the smaller

IX. FULL MODEL 177

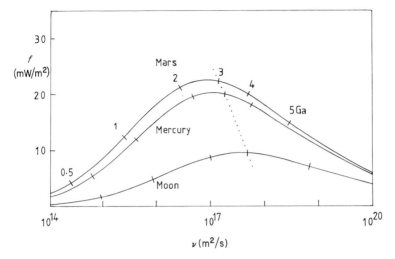

Fig. 8.7. Simplified model behaviour for Mercury, the Moon and Mars. Heat flux f (mW/m²) as a function of kinematic viscosity v (m²/s) for initial global representative temperature 2750 K and $H = 1$ MJ/kg. The duration of volcanism, in Ga, is also indicated.

bodies is, however, fairly good and although the simple model is a useful illustration, when compared with the full model it is seen to be qualitatively inadequate.

IX. FULL MODEL

The data available with which to calibrate a model, even such a rudimentary one such as that described here, is inadequate. The data we have is as follows:

1. Properties at $t = 0$ (θ_0 etc.)—no data;
2. Age of system, 5 Ga;
3. Mass, radius, mean density and surface temperature of each planet now;
4. Mercury: $r_c = 1806$ km (unreliable); volcanism ceased—no reliable estimate but heavy cratering suggests early cessation as on the Moon;
5. Venus: $r_c = 3930$ km (unreliable); observed atmospheric phenomena indicate current volcanic activity;
6. Earth: $r_c = 3470$ km; $f = 50$ mW/m²; volcanically active; source depth of volcanism of order 100 km; kinematic viscosity of upper mantle

of order $10^{17}\,m^2/s$; laboratory values of melting points and melting point gradients for basic and ultrabasic near-surface rocks;
7. Moon: $r_c = 980\,km$ (unreliable); volcanism ceased at $t = 2.0\,Ga$ (dubious estimate, discounted here, of $f = 16$–$22\,mW/m^2$ at Apollo 17, 15 sites with unreliable shallow probe method—Langseth et al., 1976); an excellent dated collection of surface rock samples;
8. Mars: $r_c = 2620\,km$ (unreliable); volcanism ceased at $t = 2$–$4\,Ga$, taken here as about $t = 3\,Ga$;
9. A very wide range of speculative and guessed data of no value.

Of this collection of data only three items, other than items (1)–(3), are reasonably accurate: the Earth's heat flux now; the Earth's core radius now; and the duration of Lunar volcanism.

A wide range of choice of model parameters allows this data to be fitted. In this work I therefore, for the purpose of illustration, choose nominal values of the model parameters. These have been selected from numerical model runs (indeed several thousand for each planet) to give what in my judgement is a reasonable compromise—undoubtedly other workers would pick somewhat different values, but for the given data and model these would, I am sure, present a similar picture.

I place emphasis on the use of the estimated core radius and duration of volcanism, rather than on the heat flux, in selecting these parameters. Also in any iterative cycle of reassessing the current choice I have looked at the terrestrial bodies in the order Moon, Earth, Mars, Mercury and Venus. In general, therefore, I have focused attention on the available data in order of presumed reliability.

In this section I sketch the model identification and provide a brief discussion in the following, final section.

Identification of Model Thermal Histories

The method used here to calibrate individual thermal histories is summarized in "identification diagrams". These display, for a given θ_0 and given material properties, numerical model results for: (1) r_c, the core radius at $t = 5\,Ga$; (2) f, the surface heat flux for the model with r_c at $t = 5\,Ga$, equal to the value given in Chapter 7; (3) the duration of volcanism, namely the value of t at which the partial melt fraction e first reaches zero everywhere in the mantle. The diagrams for the terrestrial planets all have similar form, their main feature being the sensitivity to H; they indicate a fairly narrow band of values of H compatible with the known values of r_c and volcanic duration.

The identification and behaviour of the thermal histories of the terrestrial planets is summarized in Tables 8.3–8.5 and Figs 8.8–8.17.

Table 8.3. Thermal history model parameters, and values at $t = 5$ Ga

	θ_0 (K)	f_0 (W/m²)	H (MJ/kg)	ΔE (MJ/kg)	r_c (km)	δ (km)	f (mW/m²)	z_m (km)	t' (Ga)
Mercury	2500	0.9	3.62	2.6	1806	230	15.2	—	2.2
Venus	3000	5	5.15	6.0	3930	93	43.7	58	15.7
Earth	3000	10	6.37	10.8	3470	98	50.0	107	7.1
Moon	2400	0.11	1.00	0.7	980	520	6.5	—	2.0
Mars	2500	0.5	3.00	1.4	2620	230	18.4	—	3.2

$\tau_p = 0.4$ Ga, $b = 10^4$ K, $\alpha = 1400$ K, $\beta = 6 \times 10^{-5}$ K/bar, z_m = depth to top of partial melt zone if present, t' = duration of volcanism, $\Delta E = 3 f_0 \tau_p / \rho a$, total loss per unit mass from magma convection.

Table 8.4. Energy budgets for the terrestrial planets

	E_0 (MJ/kg)	E (MJ/kg)	$c\theta_0/E_0$ (%)	H/E_0 (%)	w_0/E_0 (%)	$c(\theta_0-\theta)/\Delta E$ (%)	$H(1-r^3/a^3)/\Delta E$ (%)	$(w_0-w)/\Delta E$ (%)	mc (%)
Mercury	7.8	4.5	32	47	21	24	66	10	79
Venus	14.0	7.3	21	37	42	11	56	33	90
Earth	20.1	8.5	15	32	53	7	46	47	93
Moon	3.4	1.7	71	29	0	52	48	0	42
Mars	5.5	3.2	46	54	0	30	70	0	61

Total available energy in MJ/kg: E_0 at $t = 0$; E at $t = 5$ Ga. All other quantities are ratios expressed as percentages. Contributions to E_0 and to loss $\Delta E = (E_0 - E)$ from $c\theta_0$, H, w_0. Fraction of loss $mc \equiv 3 f_0 \tau_p / \rho a \, \Delta E$, arising from "magma" convection (the fraction of loss from "solid" convection is $1 - mc$).

8. THERMAL HISTORY

Table 8.5. Summary of gross behaviour of terrestrial planets

	r_c	δ	θ	f
Mercury	0.9	3.0	0.6	2.2
Venus	0.7	10.2	8.5	2.2
Earth	0.6	6.2	5.0	2.4
Moon	1.9	2.4	0.6	2.6
Mars	1.9	4.7	1.0	2.0

Times, in Ga, for: (1) r_c, mantle thickness reaches 0.8 of present thickness; (2) δ, upper mantle thickness twice its initial value; (3) θ, global representative temperature reaches 2150 K; (4) f, heat flux reaches twice its present value.

Mercury.

Identification diagram, Fig. 8.8

1. Curves of r_c as a function of H are drawn for $f_0 = \theta(0.5)1.5\,\text{W/m}^2$. In the region of interest, where the lines cross $r_c = 1806$ km, the curves are of small slope. The value of r_c is insensitive to the choice of H. It would not be possible to choose a particular pair (f_0, H) given only the r_c value.
2. The duration of volcanism curve is very sensitive to the choice of f_0 and the corresponding H to fit r_c. For example, $H = 3$ MJ/kg gives a duration of 6 Ga; 4 MJ/kg gives 1.1 Ga.
3. The heat flux (at $t = 5$ Ga with $r_c = 1806$ km) varies roughly inversely with H.

The available data is an uncertain core radius and unknown volcanic duration and present-day heat flow. Some coarse limits can be set. (1) Suppose $f < 20\,\text{mW/m}^2$. Then $H > 3$ MJ/kg and $f_0 > 0.7\,\text{W/m}^2$. (2) Suppose the volcanic duration was not too different from that of the Moon, say 1–3 Ga. Then $3.5\,\text{MJ/kg} < H < 4.0\,\text{MJ/kg}$ and $f_0 > 1\,\text{W/m}^2$, $f < 16\,\text{mW/m}^2$.

Model behaviour summary, Fig. 8.9

Mercury looks similar to the Moon. For the moment, therefore, I choose parameters to give a comparable duration of volcanism—the table values give 2.2 Ga.

1. $0 < t < 1$ Ga. The heat flux is high, there are strong drops in global temperature and core radius, and a rapid increase in the thickness of

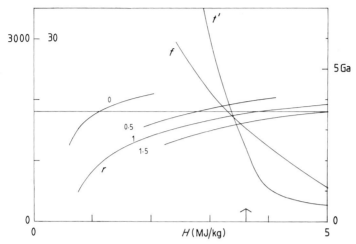

Fig. 8.8. Mercury: identification diagram. Properties as a function of H (MJ/kg) at $t = 5$ Ga: labelled curves of core radius r (km) for flux $f_0 = 0(0.5)1.5$ W/m²; heat flux f (mW/m²) for value of H at which $r_c = 1806$ km, the estimated core radius, indicated by a horizontal line; and the corresponding duration of volcanism, t' (Ga). Linear scales—the ordinate labels 3000, 30 and 5 refer to r, f and t'. Note: these scales are arranged similarly on Figs 8.8, 8.10, 8.12, 8.14 and 8.16.

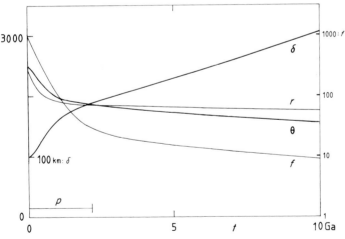

Fig. 8.9. Mercury: thermal history. Properties as a function of time t (Ga): heat flux, f (mW/m²); representative temperature, θ (K); core radius, r (km); upper mantle (sublayer) thickness, δ (km); interval of volcanism, p. Linear scales—the ordinate labels 3000, 3000 and 100 refer to θ, r and δ. The logarithmic scale refers to f. Note: these scales are arranged similarly on Figs 8.9, 8.11, 8.13, 8.15 and 8.17; the axes are the same for all quantities, except for the ordinate scale for r in Figs 8.11 and 8.13.

the upper mantle. The bulk of the global structural change occurs in this interval.
2. $1 < t < 2.2$ Ga. Rates of change are reducing and partial melting comes to an end. Heat flow is still high but rapidly falling.
3. $2.2\,\text{Ga} < t < 10\,\text{Ga}$. The planet is volcanically dead and there is only a slow change, especially small for r_c and θ—global structural changes are minor.

Venus.

Identification diagram, Fig. 8.10

1. Curves of r_c as a function of H are drawn for $f_0 = 3, 5, 7\,\text{W/m}^2$.
2. The duration of volcanism curve is very steep. For volcanic activity to be present now requires $H < 5.7\,\text{MJ/kg}$.
3. The heat flux curve is steep. A heat flux in the range 40 to 60 mW/m² requires $4.7\,\text{MJ/kg} < H < 5.3\,\text{MJ/kg}$.

Model behaviour diagram, Fig. 8.11

Values $f_0 = 5\,\text{W/m}^2$ and $H = 5.15\,\text{MJ/kg}$ give a present-day heat flux $f = 43.7\,\text{mW/m}^2$ comparable to that of Earth. The gross behaviour is similar to that of the other terrestrial planets but the following features are noteworthy.

1. There is a short interval, about 0.2 Ga, during which θ is nearly constant. There is a weak maximum of θ in this interval, owing to the role of released gravitational energy.
2. The heat flux range is great, falling from 5.16 to 0.1 W/m² in the first 2 Ga.
3. The change in δ is small and slow, especially after the first 1 Ga—δ barely doubles in 10 Ga.
4. The outstanding feature is the persistence of volcanism for a possible 15.7 Ga.

Earth.

Identification diagram, Fig. 8.12

1. Curves of r_c as a function of H are drawn for $f_0 = 8, 10, 12\,\text{W/m}^2$.
2. The curves of both volcanic duration and heat flux are very steep. For example, f of 40 to 60 mW/m² requires H of 6.25 to 6.55 MJ/kg, volcanic duration 3.4–8.3 Ga.

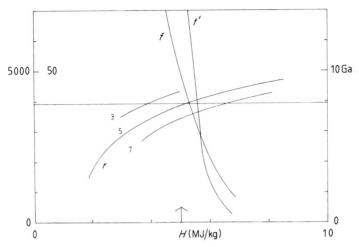

Fig. 8.10. Venus: identification diagram. (The ordinate labels 5000, 50 and 10 refer to r, f and t.)

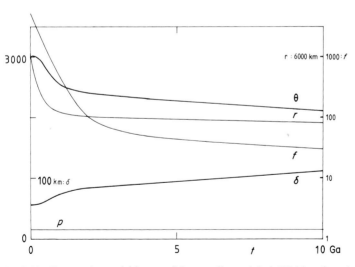

Fig. 8.11. Venus: thermal history. (Note ordinate label 6000 km for r.)

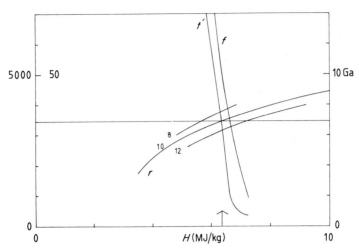

Fig. 8.12. Earth: identification diagram. (The ordinate labels 5000, 50 and 10 refer to r, f and t.)

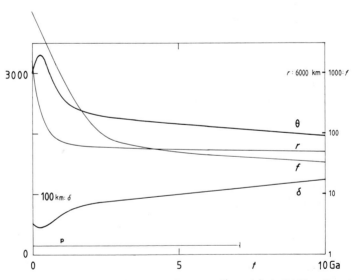

Fig. 8.13. Earth: thermal history. (Note ordinate label 6000 km for r.)

IX. FULL MODEL

Model behaviour, Fig. 8.13

There is a sharp fit to r_c, f (for the given θ_0, v_0 etc.).

The behaviour is very similar to that of Venus except that the initial behaviour is somewhat stronger, owing to the body being larger and having a greater available gravitational energy. There is a pronounced peak in θ near 0.3 Ga with a temperature rise of 300 K.

Moon.

Identification diagram, Fig. 8.14

1. Curves of r_c as a function of H are drawn for $f_0 = 0$, 0.1, 0.2 W/m².
2. The duration curve is very steep.
3. Values of f near the solution point, even for extreme choices of the parameters, are less than 10 mW/m². This is another indication that the Apollo heat flow values, based on an unreliable method, are erroneous.

Model behaviour summary, Fig. 8.15

The behaviour is grossly similar to that of the other small terrestrial planets. There are some notable differences which arise because this is the smallest body.

1. The global temperature fall is extreme.
2. The sublayer thickness is both initially high and increases rapidly—indeed it reaches mantle thickness after 9 Ga.

Mars.

Identification diagram, Fig. 8.16

1. Curves of r_c as a function of H are drawn for $f_0 = 0$, 0.5, 1 W/m².
2. The duration of volcanism curve is steep. For example, the range of H of 2.5 to 3.5 MJ/kg requires duration of 1.2–4.7 Ga with the corresponding range of heat flux f of 14 to 22 mW/m².

Model behaviour summary, Fig. 8.17

The temporal behaviour of Mars is very similar to that of Mercury. This is particularly so for θ, f and δ—except that the initial rate of fall of θ is slower for Mars, and there is a correspondingly longer interval of volcanism. The

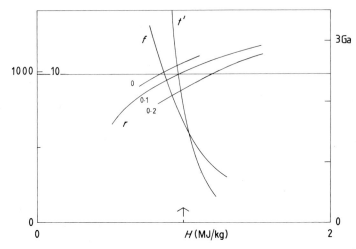

Fig. 8.14. Moon: identification diagram. (The ordinate labels 1000, 10 and 3 refer to r, f and t.)

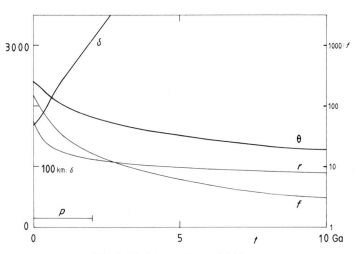

Fig. 8.15. Moon: thermal history.

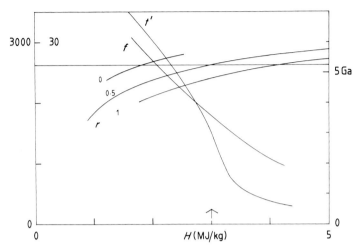

Fig. 8.16. Mars: identification diagram. (The ordinate labels 3000, 30 and 5 refer to r, f and t.)

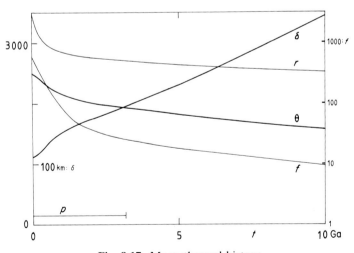

Fig. 8.17. Mars: thermal history.

values of θ, f and δ for the two bodies are also nearly the same now. (These two bodies differ in mass only by a factor of two; and the gravitational energy available for Mercury partly compensates for the effects of its smaller size.)

X. GEOLOGICAL SIGNATURE

The preoccupation of this book is the global structure of planetary bodies.

This point of view has little bearing on the detailed considerations which are the preoccupation of geology. Because of the nature of its data base, and with its ability to interpret the past using its knowledge of currently active processes, geology looks from the present progressively deeper into the past. This is moving from the known into the unknown. As we go further and further back into the past we find that the data base is thinner, and more and more fragmentary; the earliest scraps of data are unrecognizable, and then information ceases.

The point of view of this book starts in the astronomical past and looks towards the future. (I look down the other end of the telescope.) It provides a perspective for geology. This is especially so for the first 1 Ga, the interval in which the existing geological record is nearly empty and in which the bulk of the structural development occurs. Thus although the book looks at the whole 10 Ga span of possible geological time it emphasizes the initial 1 Ga interval after which the terrestrial planets are more or less structurally the same as today.

A. Energy Budget

The energy budget for the terrestrial planets is given in Table 8.4. There are a number of features to notice.

1. The initial available energy per unit mass E_0 ranges from 3.4 to 20.1 MJ/kg. This is in order of size except for Mercury, owing to its somewhat larger H than that of Mars, but mostly because of the available gravitational energy.
2. The biggest contributors to E_0 are: available gravitational energy for the most massive bodies Venus and Earth; the excess thermal energy of the core for Mercury and Mars; and the solid planet thermal energy for the Moon, the small, more nearly homogeneous body.
3. The fraction lost $\Delta E/E_0$ (not tabulated explicitly) is close to 0.5 for all the bodies ranging from 0.42 for the Moon to 0.58 for Mercury and Mars. (The system is a long way from being bankrupt.)

X. GEOLOGICAL SIGNATURE 189

4. The biggest contributors to ΔE are the same as those to E_0, except for Venus for which the core excess is the main contributor.
5. The model considers only two transport processes, "magma" and "solid" convection. Magma convection is the dominant process for all except the Moon.

This last point is one of the most striking and important results of this study of the thermal history of the terrestrial planets. The major process which establishes the thermal and physical structure is magma convection. This process is strong during the first 1 Ga but is negligibly weak now—the physical and thermal structure we find today was already established 4 Ga ago and only small changes have occurred since. This is a very different scenario from that previously envisaged, in which solid convection was the leading candidate not only for global rearrangement but also as the main heat transport mechanism throughout geological time. Solid convection remains as the leading candidate after 1 Ga of geological time. But in that initial stage it was of minor significance—magma convection was dominant.

Little is known or even conjectured on the mechanics of global magma convection. What is needed is a number of laboratory and numerical experiments and corresponding analytical studies comparable to the large body of work done for ordinary convection, in order to set up and explore models of the first 1 Ga of the geological life of the terrestrial planets, and in order to give an insight into what to look for and how to decode the fragments, as yet to be discovered, of the beginnings of geology.

B. The Time Scales

The model temporal behaviour is determined by three time scales (already referred to above):

1. τ_p, the time scale of magma convection, set at 0.4 Ga;
2. $\tau_c = \rho a H / f_0$, the time scale of the core energy loss process, and thereby of the rate of change of the core radius, with model values (in Ga): Mercury, 1.7; Venus, 1.0; Earth, 0.7; Moon, 1.7; Mars, 2.6;
3. $\tau_\theta = (A_c/A_0)^{1/3} a^2/\kappa$, the time scale of solid convection in the mantle with model values (in Ga): Mercury, 14.0; Venus, 18.8; Earth, 19.1; Moon, 10.4; Mars, 15.9.

These time scales, with $\tau_p < \tau_c < \tau_\theta$, thus play their role more or less sequentially, although in a highly non-linear system such as described here their effects are intermingled. A summary of the temporal behaviour is given in Table 8.5.

The gross geological consequence of the model presented here is

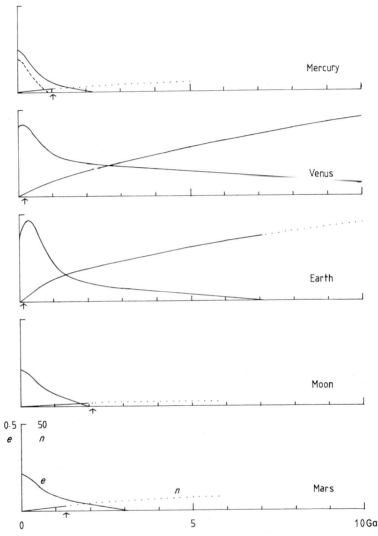

Fig. 8.18. Geological signature of the terrestrial planets. (1) The maximum partial melt fraction $e_{(max)}$ as a function of time (Ga) for selected model. (2) The number of times n the upper mantle and crust have been rearranged—calibrated to once for the present Earth in 0.3 Ga. The arrow indicates when partial melting no longer reaches to the base of the mantle.

X. GEOLOGICAL SIGNATURE

presented in the "signature" of the terrestrial planets shown in Fig. 8.18. For each planet two curves are drawn: (1) a measure of the vigour of volcanism, the maximum partial melt fraction $e_{(max)}$ (at a given time, the largest value of the radial profile of e) as described in Section VI; and (2) a measure of the vigour of convection in the mantle, the number of times the mantle has stirred itself, n, as described in Section VII.

As I now bring this work to a close I will make a few final remarks drawing on the information contained in the geological signature.

C. Remarks on Crustal Rearrangement

A ubiquitous feature of geological activity is the continual recycling of the mantle. There is, however, a striking difference in the intensity of this process between the small planets Mercury, the Moon and Mars, where it is weak with only 6, 3 and 8 recyclings, and the larger planets Venus and Earth, where it is strong with 28 and 31 recyclings. Crustal rearrangement is a major feature of the physiognomy of Venus and Earth, but is much less important for the others. I leave it to the reader to consider the plausibility of about 30 rearrangements for Earth, keeping in mind the contemporary acceptance of vigorous rearrangement during the current episode but the apparent reluctance to extrapolate this observation back into the past. The model presented here is unequivocal in its indication of the great vigour of our planet's geology throughout its geological life.

D. Remarks on Volcanism

Shield volcanoes can be represented as manometric systems in lithostatic equilibrium embedded in the crust and upper mantle (Elder, 1981). The ultimate height of a volcanic system (ignoring the role of the crust)

$$h = \left(\frac{\rho_m - \rho_a}{\rho_a}\right)H + \left(\frac{\rho_a - \rho_b}{\rho_a}\right)h_*$$

where h is the volcanic height about the paleo-surface, H is the source depth and h_* is the extent of the magma column which is composed of basaltic magma. The various densities are: ρ_m, mantle; ρ_a, ultrabasic magma; ρ_b, basaltic magma. The column at depth is assumed to be filled with a hypothetical ultrabasic magma derived from the mantle and which at a particular level differentiates to produce a basaltic magma. Various studies suggest that the differentiation occurs at a pressure of order 3 kbar, namely at a depth in the magma column of order 10 km.

We cannot be certain of the values of the various quantities but since the relation provides a linear relation of h and H, and the h_* term is relatively small, we can use the relationship with some confidence after calibrating it for terrestrial volcanoes. This is equivalent to choosing the densities. Our best estimate for h on Earth is provided by Hawaii for which $h \approx 9$ km. Thus with ρ_m, ρ_a and ρ_b taken as 3300, 3100 and 2800 kg/m^3:

$$h = (2H + 3h_*)/31.$$

For $h_* = 10$ km and $H = 120$ km, then $h = 9$ km. Note that here H is taken as the global value for the present Earth.

The corresponding model value for Mars at which volcanism ceases is $H = 336$ km, for which the relation gives $h = 22.6$ km. This is close to, if a little larger than, the height of Olympus Mons at about 20 km. The model values are of course nothing more than rough estimates but the agreement is certainly of the correct order of magnitude and suggests that this idea provides a useful calibration of the thermal history model.

This identification needs to be treated with some caution on other grounds. Consider the problem of estimating the temporal variation of the global rate of production of surface volcanics. For example, if a volcanic system affects an area $\sim H^2$, then the number of systems $\sim (a/H)^2$; the net permeability of the conduit decreases with pressure so that the net permeability $\sim H^{-n}$, $n > 0$; and noting that $h \sim H$, the global rate $\sim H^{-(2+n)}$. This suggests that volcanism diminished strongly with geological time—so that volcanism near the terminal state of global volcanism would be a rare event and what we see, for example today on Mars, is a relic of an earlier time. (The model values of H at $t = 0, 1, 2$ and 3 Ga give $h = 6, 11, 14$ and 19 km.)

In isolation, this is a plausible description for Mars. It is plainly, however, not so for Mercury and the Moon, both of which have had a vigorous volcanic history, but which have volcanoes of heights of at most a few kilometres. There is no obvious modification of the manometer model which could account for the small heights—indeed at the cessation of volcanism, in the quoted models, the depth to partial melting is 330 and 610 km, respectively. Some other effect must have operated.

Inspection of the Moon model, over a wide range of parameters compatible with the presumed known radius of the core and the duration of volcanism, shows that, in contrast to Venus, Earth and Mars, the zone of partial melting, throughout the volcanic history, extends down to the mantle–core boundary. This is a situation different from that of a localized zone of partial melting bounded both above and below by solid mantle. The essential feature of the manometer model of volcanism is the difference between the pressure in the surrounding solid mantle and that in the liquid column—it is this pressure difference, lithostatic minus "magmastatic", which drives the system and provides the head to elevate magma above the paleo-surface. Where, however, the entire mantle is partially molten (and even if the distribution of melt is patchy) the ambient pressure will be closer to "magmastatic" than lithostatic. The available head to drive a volcanic system will be reduced.

X. GEOLOGICAL SIGNATURE

Shallow discharges will be readily possible—indeed would be the most common eruptive mode—but high volcanoes would not. Small shield volcanoes would be possible in a region of marginal partial melt. Here, then, is the other effect—and the data on duration and core radius is sufficiently reliable for a plausible case to be made that volcanism on the Moon occurred with partial melting extending throughout the mantle for the entire volcanic stage.

There remains the enigma of Mercury. In the model presented above, largely on the visible evidence that Mercury is cratered similar to and at least as much as the Moon, parameters were chosen to give a volcanic duration of 2.2 Ga. This gives a pattern of volcanism similar to that of Mars—the first half of the volcanic stage has partial melting throughout the mantle but during the latter interval partial melting is confined to a localized zone bounded above and below by solid mantle. This suggests that the model chosen for Mercury needs to be modified. The data shown in Fig. 8.8 for Mercury includes that for $f_0 = 1$ W/m² and $H = 3.8$ MJ/kg, for which volcanism is active for 0.9 Ga with mantle-wide partial melting for 0.8 Ga. (Incidentally this shows the sensitivity of certain aspects of the model behaviour to parameter variation.)

This is possibly an extreme choice—the bulk of the surface obliteration by cratering is over in the first 1 Ga (see Fig. 2.4). A repeat of the Apollo programme on Mercury (and Mars too, please!) will be necessary to resolve questions of this kind.

This early part of the volcanic stage during which partial melting extends to the base of the mantle will occur for all the terrestrial planets. The selected model results give the following values for the duration and mantle thickness at its close:

Mercury: 1.0 Ga, 520 km (alternatively, see above, 0.8 Ga, 500 km);
Venus: 0.2 Ga, 730 km;
Earth: 0.1 Ga, 650 km;
Moon: 2.0 Ga, 600 km;
Mars: 1.3 Ga, 580 km.

During this interval, by analogy with the Moon, surface volcanism takes the form of shallow, extensive flows of hot, low-viscosity lavas with minor central volcanism, and that of small height. On Earth and Venus it is extremely unlikely that any relics of this early volcanism remain, following the obliteration produced by a continuing vigorous geology, whereas this process has formed the preserved lunar volcanic surface and is probably a major contributor to the preserved volcanic surface of Mercury and Mars.

E. Geological Style

All the terrestrial planets start their geological lives with great vigour. The Earth itself would have been an awesome place, with frequent volcanism and endemic intense hydrothermal and phreatic activity (such as seen today in small restricted geothermal areas on land and near ocean ridges—the term "Hadean" era, used by some authors for an early interval of thermally intense geological activity, is certainly apt). There will however be little preserved from that time. The bulk of the possible evidence has been obliterated by cratering, flood volcanism or many cycles of crustal reworking.

During its geological life a terrestrial planet passes through three stages determined by the partial melt structure of the mantle.

1. Mantle-wide partial melting. This early stage occurs for all the terrestrial planets. It is characterized by vigorous sheet volcanism. Crustal rearrangement also occurs, possibly with very small regions of coherent crustal displacement. The Moon and possibly Mercury spend their entire volcanic history in this stage. These are the most primitive bodies with the least developed crustal systems. This stage is very brief for the large terrestrials Venus and Earth.
2. Restricted vertical zone of partial melting. This next stage is characterized by both sheet and central volcanism, together with crustal rearrangement, of the style found today on Earth and presumably on Venus—these two spend the bulk of their volcanic history in this stage. The vigour of the mantles of these two bodies leads to a highly evolved crustal system.
3. No partial melting. In this stage "solid" convection continues in the mantle but there is no volcanism. Crustal rearrangement will be dominated by rifting and folding. This stage is clearly seen today on Mars.

    (For the Moon and Mercury there is some post-volcanic rifting and faulting visible in the surface topography but it is muted, indicating a feeble mantle after volcanism ceased. Since the end of volcanism until now, the model values for the number of rearrangements are: Moon, 1.0; Mars, 1.9; and Mercury, 2.2.) From this point of view Mars is the most developed of the terrestrial planets—it has spent time in each stage and has passed into this terminal structural stage with its mantle still (weakly) active.

Once partial melting can no longer occur a key factor in the chain of processes rearranging the crust and upper mantle is broken. Vigorous geological activity is no longer possible. The body reaches its geological old age. With the exception of Venus, this is the fate of all the terrestrial planets. The mantle will continue to grow but no further global phase changes will occur until the dispersion of all the planets when the Sun rapidly expands as it begins its sub-giant stage. The record will be wiped clean—not even the green black rock will remain.

Appendix

I. POLYTROPIC MODELS

The material of this section is a mere outline of the subject treated in more detail in numerous articles and books: e.g. those by Novotny (1973) or Clayton (1968). The original work was done by Emden (1907).

Many self-gravitating bodies can be represented as model systems close to hydrostatic equilibrium. Where also the equation of state has the simple form $P/\rho' = $ constant, that of a so-called polytrope, a very simple model results. The quantity $n = 1/(\gamma - 1)$ is known as the polytropic index. For an adiabatic gas $\gamma = C_p/C_v$ which for an ideal monatomic gas is 5/3 ($N = 3/2$). The case $\gamma = 4/3$ ($n = 3$) was noted originally by Eddington to provide a fair model of a star. (In this section only, γ and n are interchanged from the text usage to that more commonly used.)

For a spherical body in hydrostatic equilibrium composed of a gas in adiabatic equilibrium:

$$dP/dr = -g\rho$$
$$g = GM/r^2$$
$$dM/dr = 4\pi r^2 \rho$$
$$P = K\rho^\gamma$$

where M is the mass enclosed within radius r. Writing

$$\gamma = (n+1)/n$$
$$\rho = \lambda \phi^n$$
$$r = a\xi$$

with λ and a the (Emden) scales of density and length, and eliminating P, we obtain the Emden equation:

$$\frac{1}{\xi^2}\frac{d}{d\xi}\left(\xi^2\frac{d\phi}{d\xi}\right) = -\phi^n$$

provided

$$a^2 = (n+1)K\lambda^{(1-n)/n}/4\pi G.$$

where the quantity a is to be determined from the physical conditions of the model.

The boundary conditions, normally of interest, are arranged as follows:

1. At $\xi = 0$. This is the centre of the body, $r = 0$, and since $dP/dr = 0$ then $d\phi/dr = 0$. Customarily we choose $\lambda = \rho_c$, the central density, so that $\phi = 1$.
2. At $\xi = \xi_1$. For $n < 5$ the first zero of ϕ occurs at $\xi = \xi_1$, say. This corresponds to $\rho = 0$, $P = 0$. Thus the "surface" of the body is taken as a distinct level of nominal zero density. This is not strictly realistic but in practice provides a good approximation to the location of, say, the photosphere.

For an ideal gas: $T/T_c = \phi$, $P/P_c = \phi^{n+1}$. There are a number of quantities of interest:

1. Quantities at the centre, $\xi = 0$:
   density, λ (by definition);
   pressure, $P_c = (GM^2/R^4)/4\pi(n+1)(d\phi/d\xi)_1^2$;
   temperature, $T_c = \beta P_c/\lambda$.
2. The mass distribution

$$M(\xi) = \int_0^r 4\pi r^2 \rho \, dr = 4\pi a^3 \lambda \left(-\xi^2 \frac{d\phi}{d\xi}\right)$$

Hence the ratio of mean to central density is

$$\bar{\rho}/\rho_c = -3(d\phi/d\xi)_1/\xi_1$$

In practice we are interested in the identification of a particular case. For example, consider the case of a given mass M and radius R.

1. From M, R find: $\bar{\rho} = M/\frac{4}{3}\pi R^3$, then $\lambda = \bar{\rho}/[-3\phi'/\xi]_1$.
2. From R find: $a = R/\xi_1$.

I. POLYTROPIC MODELS

3. With λ, a then find K from $a^2 = (n+1)K\lambda^{(1-n)/n}/4\pi G$.
4. Then for r such that $0 \leq r/a \leq \xi_1$:

$$\rho = \lambda \phi^n$$
$$P = K\rho^{(n+1)/n}$$
$$T = \beta P/\rho$$

where $\beta = \mu m_H/k$, μ = mean atomic mass, $m_H = 1$ amu, k = Boltzmann's constant and $m_H/k = 1.21 \times 10^{-4}$ kgK/J. In particular: $\rho_c = \lambda$, $P_c = K\lambda^\gamma$, $T_c = \beta P_c/\lambda$.

A. Polytropic Envelope

The Emden equation applies to the envelope of gas with the proviso that the scales ρ_c, P_c and T_c are now mere scaling parameters (and are not the values at $r = 0$) although ϕ is still a dimensionless temperature. The model is identified as follows:

1. $a = R_g/\xi_1$, where ξ_1 is the zero of ϕ, and hence $\xi_2 = R_\ell/a$, where R_g and R_ℓ are the outer and inner radii of the envelope.
2. Writing $y = -\xi^2 \, d\phi/d\xi$, $M_g = 4\pi a^3 \lambda(y_1 - y_2)$ which can be solved for λ. K is then found from the expression for a^2. All other quantities can then be determined.

B. Polytropic Sheet

The internal structure of a thin, disc-like body can be modelled as a sheet whose properties are independent of lateral position. A similar set of relations to those for a sphere applies except that: the coordinate r is the distance measured, from the central plane, transverse to the sheet; $g = 4\pi Gm$, where $m(r)$ is the mass per unit area between the central plane and level r; and $dm/dr = \rho$. The transformation and the related relations are unaltered. The equation for ϕ becomes $\phi'' = -\phi^n$. As before, ϕ is an even function; with, for $n = 3/2$, $\phi \approx 1 - \frac{1}{2}\xi^2 + \frac{1}{16}\xi^4 - 0.0053\xi^6$ giving a fit with error less than 0.001; and with its first zero at $\xi_1 = 1.6$, $\phi'_1 = -0.9$. The total mass per unit area $m_1 = a\lambda(-\phi'_1)$, so that the mean density $\bar{\rho} = \lambda(-\phi'_1)/2\xi_1$. The surface temperature gradient $(dT/dr)_1 = 1.5T_c/\Delta$, where Δ is the half-thickness of the sheet.

C. Quasi-Polytrope

Whereas for a gaseous body a polytropic model with $P/\rho^\gamma =$ constant is often a good first approximation, for stony bodies we find $(\rho/\rho_0)^\gamma = 1 + \gamma\chi_0 P$

is appropriate, where χ_0 is the compressibility at zero pressure. A body with this P, ρ relationship could be named a quasi-polytrope.

The description of a self-gravitating quasi-polytropic sphere proceeds in an identical manner to that of a polytrope and we obtain the same Emden equation—except that the variable ϕ is no longer equal to T/T_c.

The identification of a particular model proceeds somewhat differently, since the surface boundary has $\rho = \rho_0$, some given non-zero surface density (rather than $\rho = 0$). Consider being given n, ρ_0, R, M.

1. At the surface $\xi = \xi_s$, $\rho = \rho_0 = \rho_c \phi_s^n$.
2. The total mass from the definition and the Emden equation $M = 4\pi a^3 \rho_c(-\xi_s^2 \phi_s')$. Hence putting $\rho_c = \rho_0/\phi_s^n$ and noting that also $M = \frac{4}{3}\pi R^3 \bar{\rho} = \frac{4}{3}\pi a^3 \xi_s^3 \bar{\rho}$, then

$$\bar{\rho}/\rho_0 = (-3\phi'/\xi\phi^n)_s$$

This can be readily solved (for example by interpolation using a table of ϕ) for ξ_s.
3. Then we have $a = R/\xi_s$. All other quantities can now be easily worked out.

II. LUMINOSITY

A major item required in evolutionary models is the rate at which energy can be lost from the body. This is controlled by the internal temperature distribution and by the opacity of the material. The net power loss L is usually represented by referring to the temperature T_s at the photosphere, near an optical depth of unity, as $L = 4\pi\sigma R^2 T_s^4$. Our object is then to find T_s.

A. Opacity

The radiation intensity (W/m^2) loss in passing through a partially transparent medium is usually represented by

$$dI/I = -\kappa\rho\, dr = -d\tau$$

where κ is the mass absorption coefficient and τ the optical depth. The absorption is a very detailed function of the wavelength of the radiation. Where these details are not of interest an appropriate mean absorption, most commonly the "Rosseland mean", is used. The form of $\kappa = \kappa(\rho, T)$ for a solar mixture is shown in Fig. A1 (from data by Cox and Stewart, 1970,

II. LUMINOSITY

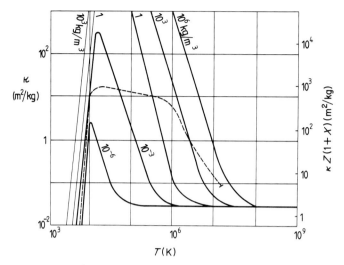

Fig. A1. Opacity κ (m²/kg) as a function of temperature T (K) for various densities (kg/m³) for material of approximately solar composition with $X=0.80$, $Y=0.19$ and $Z=0.01$. The dashed line is for the radial opacity profile of a model of the Sun.

and given also by Novotny, 1973, pp. 510–511). There are three distinct zones in this diagram.

1. $T \lesssim 10^4$ K. κ is very strongly dependent on T. This is the zone of ionization. Very roughly $\kappa/\rho^{1/2}T^4 \approx \text{constant} = 2 \times 10^{-12}Z(1+X)$.
2. $T \gtrsim 10^4$ K. κ is dependent on both ρ and T. This is a zone of fully ionized gas, largely a gas of protons and electrons. Very roughly $\kappa/\rho T^{-7/2} \approx \text{constant} = 5 \times 10^{23}Z(1+X)$.
3. For a given ρ at sufficiently high T, κ tends to a constant value, about $0.03 \text{ m}^2/\text{kg}$.

B. The Photosphere

The luminosity from the interior

$$L = -\frac{64\pi\sigma r^2 T^3}{3\kappa\rho}\left(\frac{dT}{dr}\right)_s$$

where for a polytropic sphere

$$\left(\frac{dT}{dr}\right)_s \approx -0.74 T_c/R$$

If the incident solar flux (W/m^2) is s the photosphere temperature T_s satisfies

$$4\pi R^2 \sigma T_s^4 = L + \pi R^2 s$$

This equation is readily solved iteratively for T_s.

III. THE ESCAPE FLUX

A detailed assessment of the escape flux is very complicated (see, e.g., Jeans, 1925; Chamberlain, 1978, Chapter 7). In this work I have made the following assumptions.

1. Escape occurs from an "exo-base" at a height above the "photosphere" (the level at which the optical depth ≈ 1) given by $z = \zeta H$, where $H = T_s/\beta g$ is the pressure scale height evaluated at the photosphere at photosphere temperature T_s. Note that the scale heights for pressure, density and temperature are H, γH and $\gamma H/(\gamma - 1)$. The phenomenological ratio $\zeta = 2.8$ in this work.
2. The mass flux is limited solely by the criterion usually referred to as the Jean's escape mechanism and not by diffusion. For a Maxwellian velocity distribution, with radial and transverse components (v_r, v_ϕ), integrating with the condition $v_\phi^2 > v_{(escape)}^2 - v_r^2$ gives the particle flux

$$f = \frac{nU}{2\pi^{1/2}} (1 + w) e^{-w}$$

where $w = (v_{(escape)}/U)^2$; $U = (2kT/m)^{1/2}$ and U is the most probable molecular speed.

References

In addition to and including the works referred to in the text and listed below, I have found the following works invaluable: Moore and Hunt (1983)—outstanding!; Henbest and Marten (1983); Weast (1983); Novotny (1973); Carmichael et al. (1974); and Cook (1980).

Allen, C. W. (1983). "Astrophysical Quantities." Athlone Press, London.
Baldwin, R. B. (1963). "The Measure of the Moon." University of Chicago Press, Chicago.
Brown, G. C. et al. (1977). The Moon. Phil. Trans. Roy. Soc. A, 285.
Bullen, K. E. (1975). "The Earth's Density." Chapman & Hall, London.
Bullen, K. E. and Haddon, R. A. W. (1967). Phys. Earth Plan. Int. 1, 1–13. See also Proc. Nat. Acad. Sci., Wash. 58, 846–852 (referred to in Bullen, 1975, p. 173).
Carmichael, I. S. E., Turner, F. J. and Verhoogen, J. (1974). "Igneous Petrology." McGraw-Hill, New York.
Carslaw, H. S. and Jaeger, J. C. (1959). "Conduction of Heat in Solids." Oxford University Press, Oxford.
Chamberlain, J. W. (1978). "Theory of Planetary Atmospheres." Academic Press, London.
Clayton, D. D. (1968). "Principles of Stellar Evolution and Nucleosynthesis." McGraw-Hill, New York.
Cook, A. H. (1980). "Interiors of the Planets." Cambridge University Press, Cambridge.
Cox, A. N. and Stewart, J. N. (1970). *The Astrophysical Journal* **19**, 243.
Descartes, R. (1644). "Principia Philosphiae."
Elder, J. W. (1976). "The Bowels of the Earth." Oxford University Press, Oxford.
Elder, J. W. (1981). "Geothermal Systems." Academic Press, London.
Emden, R. (1907). "Gaskugeln." B. G. Teubner, Leipzig.

Henbest, N. and Marten, M. (1983). "The New Astronomy." Cambridge University Press, Cambridge.
Hobbs, P. V. (1974). "Ice Physics." Oxford University Press, Oxford.
Hutchinson, R. (1983). "The Search for Our Beginning." Oxford University Press, Oxford.
Jeans, J. H. (1925). "The Dynamical Theory of Gases." Cambridge University Press, Cambridge.
Jeffreys, H. (1976). "The Earth: Its Origin, History and Physical Constitution" (6th Edition). Cambridge University Press, Cambridge.
Kant, I. (1755). "Allgemeine Naturgeschichte und Theorie des Himmels."
Lamb, H. (1945). "Hydrodynamics." Dover Publications, New York.
Langseth, M. G., Keihm, S. J. and Peters, K. (1976). *EOS Trans. AGU* **57**, 271.
Laplace, P. S. (1796). "Exposition du système du Monde."
Mestel, L. (1977). *In* "Star Formation," IAU Symposium No. 75 (T. de Jong and A. Maeder, eds), p. 213. Dordrecht, Reidel.
Moore, P. and Hunt, G. (1983). "The Atlas of the Solar System." Mitchell Beazley, London.
Mörner, N. A. (ed.) (1980). "Earth Rheology, Isostasy and Eustasy." Wiley, Chichester.
Neukum, G. (1977). Lunar Cratering. *Phil. Trans. R. Soc. Lond.* **A**, 285, 267–272.
Novotny, E. (1973). "Introduction to Stellar Atmospheres and Interiors." Oxford University Press, Oxford.
Oort, J. H. (1950). *Bull. Astron. Inst. Neth.* **2**, 91–110.
Öpik, E. J. (1963). *Geophys. J.* **7**, 490–509.
Shannon, R. D. and Prewitt, C. T. (1969). *Acta Cryst.* **B25**, 925–946.
Weast, R. C. (ed.) (1983). "CRC Handbook of Chemistry and Physics." CRC Press, Boca Raton, Florida.
Wedepohl, K. H. (1971). "Geochemistry." Holt, Rinehart & Winston, New York.
White, I. G. (1967). *Earth Planet. Sci. Letters* **3**, 11–118.

Index

Abundances, 6, 53
Accretion, 13
  hypothesis, 30
Air, 87
Allende meteorite, 11
Alloy matrix, 18
Ammonia, properties, 97
Andesite, 109
Anelasticity, 4
Angular momentum, distribution, 6, 9, 26
Apollo mission, 30
Asteroids, 30, 154
  icy, 68
Atmosphere-crust, 85
  Earth diagram, 87
  Earth development, 90
Atomic abundances, 6, 7, 53

Background, mass ratio, 26, 27
Basalt, 109, 166, 168
Binary stars, 10
Budding, 27
Budget
  energy, 178, 179, 188
  solar system, 5–6
Bullen–Jeffreys model, 134

Calcite, 88
Callisto, 2, 76
Carbon dioxide, 87
  amount, 88
  crustal, 89
  distribution, 90
  oceanic, 89
  properties, 97, 102
  sediments on Mars, 102

Carbonaceous chondrites, 11
Catastrophe, 9
Central body emergence, 70
Central mass fraction, 122
Central tendency, 26
Ceres, 2, 31, 154
Chemical development
  Earth, 116, 117, 127
  parameters, 126
  small bodies, 65–68
  terrestrial planets, 107
Chondrites, 11
  carbonaceous, 11
Chondrules, of slag, 12
Chronology, early solar system, 76
CIPW scheme, 111
Collapse ratios, 54–55
Collector model, 33
Comets, 30
  material, 2
Composition, role, 2
Compressibility, 2, 134, 141
Condensation, of nebula, 9
Contraction,
  gravitational, 38
  nebula, 22
Convection, 82, 84, 158, 164, 166
  magma, 168
  solid, 169
Core, 14, *see also* Inner core
  energy, specific excess, 174, 178
  formation, 68–70, 72–75
  formation temperatures, 72–73
  growth, 72, 75
  liquid, 13
  mass limit, 73
  material, 4
  parameter identification, 136–137
  radius, 162, 174, 178
  zero pressure density, 137

Cratering,
  amount, 32
  chronology, Moon, 34
  density, 30
  rate, Moon, 34
  shape, 32
Critical packing level, 119
Crust, 14, 85, 91–92, 95
  global average, 109
  rearrangement, 172, 191, 194
  upper continental, 109
Cumulate, 12, 123

Deformation, 4
Degassing, *see also* Depletion
  early solar, 63–64
  scenario, 68
  small bodies, 65–68
  model parameters, 61
  proto-terrestrial planets, 52, 61–63
  times, Jovians, 63–65
Density
  contrast, mantle–core boundary, 120
  development, Earth, 121
  distribution, 117
  distribution, nebula, 26
  distribution sequence, Earth, 130–131
  rock-substance, function of pressure, 134–138
  oxygen ion, 118
  Earth layer parameters, 137–138
  model identification, 140
  proto solar system, 27
  planets, 2
  rock-substance, 117
  structure, model relations, 138
  structure, large ice moons, 154–155
  structure, Io, Europa, 155–156
  zero pressure, 118, 122
  zero pressure, Earth, 137
Depletion, *see* Degassing
  from small bodies, 67
  in freckle field, 68
Depth, optical, 198
Dione, 2, 153
Disc, 22
  depletion times, 58–61
  emergence, 22–30

Disc—*contd.*
  energy budget, proto-Terrestrial planets, 57
  front, 26
  Jovian planets, 44–47
  Jupiter, 47–48
  mass ratio, 26
  model, 37, 39
  photosphere temperatures, 46
  solar, 37, 43–44
Dispersion, in shear flow, 21
Dolomite, 88
Duration of volcanism, 176, 178
Dust, 9, 30

Earth
  atmosphere, 81, 90, 97
  atmosphere-crust, 92–94
  carbon dioxide amount, 88
  chronology, 76
  core mass, 75
  core onset, 73
  core top, 164
  crustal rearrangement, 172
  density, 2
  density structure, 143, 146, 149
  emergence, 71
  energy, 161, 188
  flushing, 61
  fractionation factors, 109
  global information, 177
  heat flow, mean, 162
  layers, properties, 137
  magma convection, 169
  mantle recyclings, 191
  mass loss, time scale, 61
  molecular escape, 57
  oldest rocks, 11
  oxide ratios, 109
  partial melt development, 165–168
  physical structure now, 134
  physical structure development, 146
  pool model, 127, 129
  radial distributions, 135
  radiogenic powder scale, 173
  style, 194
  surface heat flux, 169
  temperature, global, 164
  thermal history, 176, 180

Earth—*contd.*
 thermal time scales, 189
 upper mantle energy budget, 170
 volcanism duration, 191
 water amount, 88
Earth-like bodies, 4
Ejecta, 32
Elements, abundance, 6–7
Enceladus, 2, 153
Energy
 budget, 8
 budget, proto-terrestrial discs, 71–72
 budget, thermal history, 159, 161, 178–179
 chemical, 8
 equation, for disc, 40
 gravitational, 8, 160
 kinetic, 8
 nuclear, 8
 potential, 39
 rotational, 8
 thermal, 39, 159
 thermal, core, 160
Envelope, gaseous, 14
Equation of state, Earth materials, 134
Erosion, 91
Escape
 from disc, 55–56
 from Jovian moons, 56
 mechanism, 58
 simple model, 55
 velocity, 55
Eucrites, 12
Europa, 2, 76, 155–156
Exo-base, 200
Exotic material, 2, 10

Finite strength, 4
Fluctuations, temperature, 167
Flushing, 13, 61
Fluvial processes, Mars, 102–104
Flux
 escape, 200
 heat, surface, 162, 173–174, 178
 mass, 57–58
Folding, 194
Fractionation
 evidence, 107
 function, 84

Fractionation—*contd.*
 paths, 115–116
 pre-crustal, 82–85
 ratios, Earth upper mantle, 108–110
 ratios, Earth inner core, 123–125
Freckle, 67
Fusion, 8

Galaxy, 1
Ganymede, 2, 76, 154
Gaseous body, 2, 4
Geological
 signature, 188
 style, 193
 time, start, 81, 91
Gestation, 19
Granitic material, limiting amount, 92
Gravitational,
 system, 1
 work function, 160–161

Hawaii, 5, 192
Heat
 flux, surface, 162, 173–174
 transfer, 168
Helium, 6, 59, 64–65
Henry's relation, 89
Herschel crater, 5
Heterogeneity, chemical, 107
Hydrogen, 6, 59–67
 bodies, evolution, 37
 era, evidence, 11
Hydrothermal,
 activity, early crust, 93–96
 pulse, 95
 systems, 169
Hyperion, 2, 5, 153
Hypersthene, 111–113, 126–130

Iapetus, 2, 153
Ice
 asteroids, 68
 compressibility, 152
 crust, Titan, 105
 bodies, 2–4
 phase diagram, 4, 152
 properties, 152
Imilac meteorite, 18

Inner core, 14, 164, *see also* Core
  fractionation factors, 124
  model results, 125
Interstellar material, 20
Io, 2, 76, 155–156
Ionic radii, 119
Iron meteorites, 12

Jeans mechanism, 58, 200
Jeffreys–Bullen model, 134
Jovian planets, 2
  evolution, 44
  mass loss, 63
  molecular escape, 56
  moments of inertia, 48
  moons, 61, 151–156
  moons, occurrence, 68
  onset, 30
  physical structure, 47
Jupiter
  chronology, 76
  density, 2
  depletion times, 65
  development, 47
  disc model, 45–46
  emergence, 22, 27–30
  mass loss, 63–65
  molecular escape, 57
  moons, cratering, 33
  physical structure, 48

Kepler, speed, 6, 27
Kelvin, time scale, 8
Kimberlite, 80, 168
Kinematic viscosity, 171
Kirkwood gaps, 31

Lava flood plain, ii
Limiting stress, 5
Limestone, 88
Luminosity, 8
Lherzolite, 109–110
Liquid core, 13, 72–75
Liquid
  phases, surface occurrence, 96–98
  water (temporary), Mars, 103
Loess, 102

Luminosity, 198
  Jovian planets, 47

Magma,
  basaltic, 166
  convection, 168, 178, 189
Main sequence, 38
Mantle, 14
  heat transfer, 162
  layers, 4
Mars
  atmosphere, 81, 97
  atmosphere–crust model, 100, 102–104
  carbon dioxide amount, 100
  chronology, 76
  core mass, 75
  core onset, 73
  cratering, 33
  crustal rearrangement, 172
  degassing, example, 68
  density, 2
  density structure, 143–144, 146, 147, 151
  emergence, 70
  energy, 161, 188
  flushing, 61
  global information, 177
  heat flow, mean, 162
  inner core, 125
  mantle recyclings, 191
  molecular escape, 57
  permafrost, 100
  physical structure, 143–144
  physical structure development, 151
  polar caps, 101
  pool model, 128–130
  structure during core formation, 74
  style, 194
  surface temperature, 101
  temperature, global, 164
  Tharsis region, ii
  thermal history, 174–177, 185–188
  thermal time scales, 189
  volcanism ceased, 166
  volcanism duration, 192
  water, amount, 100, 103
Mass
  central, 26

INDEX 207

Mass—*contd.*
 critical, 53, 59
 distribution, 9
 flux, disc, 57–58
 loss, 8
 particle, 31
 planets, 6
 proto-planets, 53
 Sun, 6
Mass-size relation, 2
Melting
 partial, 165
 point, 82, 166
Mercury
 chronology, 76
 core mass, 75
 core onset, 73
 cratering, 33
 crustal rearrangement, 172
 density, 2
 density structure, 144–145, 147, 149
 emergence, 70
 energy, 161, 188
 flushing, 61–63
 global information, 177
 heat flow, mean, 162
 mantle recyclings, 191
 molecular escape, 57
 physical structure, 144–145, 147, 149
 physical structure development, 148–149
 pool model, 125–126, 129
 style, 194
 temperature, global, 164
 thermal history, 176–181
 thermal time scales, 189
 volcanism ceased, 166
 volcanism duration, 191
Meteorite
 Allende, 11
 falls, 18
 Imilac, 18
 material, 11
 stony-irons, 18
 texture, 11
Meteoroid, 20
Methane
 ocean, Titan, 105
 properties, 97
Mimas, 2, 5, 153

Mineral
 assemblage, 111, 129
 distributions, 126–128
 normative, 111
Molecular
 clouds, 20
 escape, 56
 speed, 55
Moment of inertia, 6, 139–140, 148
 Jovians, 48
Moon
 chronology, 76
 comparison with ice-rock moons, 154
 core mass, 75
 core onset, 73
 cratering, 33
 crustal rearrangement, 172
 density, 2
 density structure, 145–147
 emergence, 70
 energy, 161, 188
 flushing, 61
 global information, 178
 heat flow, mean, 162
 highland rock samples, 36
 inner core, 125
 mantle recyclings, 191
 molecular escape, 57
 physical structure development, 145–148
 pool model, 128–130
 style, 194
 temperature, global, 164
 thermal history, 177–180, 185–186
 thermal time scales, 189
 volcanism ceased, 166
 volcanism duration, 191
Moons
 mass-size relation, 3
 ice-rock, 153–156
 Jovian, 61, 151–156
 Saturn, 153

Nebula, 9, 22
 hypothesis, 19
Neptune
 chronology, 36
 density, 2
 depletion times, 65

INDEX

Neptune—*contd.*
  disc model, 45–47
  emergence, 22, 27–30
  mass loss, 63–65
  molecular escape, 57
  moons, 97
  physical structure, 48
Nitrogen, properties, 97
Norm, 111
Normative minerals, 111–113
Nuclear
  energy, 6
  reaction, 5
Nusselt number, 95, 170

Oberon, 76
Obliteration, 22, 32
  Moon surface, 36
Ocean, depth, 90, 92
Olivine, 12, 18, 80, 111, 125–130
Olympus Mons, 192
Oort cloud, 2
Opacity, 198
Optical depth, 198
Origin, solar system, 9
Orion
  OB1 association, 10
  nebula, 21
  spur, 1, 12
Outline (of book), 15
Oxide ratios, Earth rocks, 109
Oxygen, properties, 97

Packing, 118–120
  critical level, 119
  random, 120
Pallas, 2, 31, 154
Pallasite, 12, 18
Partial melt, 165
  depth, 168, 178
  development, Earth, 167
Partial pressures
  carbon dioxide, 90, 92
  water, 90, 92
Particle
  collector data, 34–35
  masses, 34
Parturition, proto-terrestrial planets, 51

Peridotite, 81, 129
Perseus arm, 1
Phase
  changes, 13
  diagram, ice, 4
Photosphere, 41, 199
  temperatures, 45, 59, 85
Physical development, Terrestrial, 133
Placenta, 81
Planetismals, 9
  hypothesis, 20, 30
Planets, 2
  emergence, 27–30
  mass-size relation, 3
  terrestrial total mass, 35
Plasma, 13
Pluto, 2, 154
Poisson's ratio, 123
Polytropic
  body, 3
  envelope, 197
  index, 195
  model, 48, 195
  quasi, 197
  sheet, 197
  shell, 72
Pool, 14, 52
  model, 113
Population II stars, 63
Populations, sub-planetary material, 31
Prandtl number, 170
Prelude, 20
PPI chain, 8
Pressure scale, 2
Pyroxene, 12, 126–130

Radial distribution, Earth, 134
Radioactivity
  dating, 11
  heating, 172
Random packing, 120
Rayleigh number, 170–171
Recirculation, 172
  ocean water, 89
Recycling, mantle, 172
Resources
  atomic, 6
  energy, 8
  mechanical, 8

INDEX

Rhea, 2, 153
Rifting, 194
Rock
 oldest, Earth, 11
 primary material, 108
 solar composition, 112
Rotundity, 4

Satellite system, 2
Saturn
 chronology, 76
 density, 2
 depletion times, 65
 disc model, 45–46
 emergence, 22–30
 mass loss, 63–65
 molecular escape, 57
 moons, cratering, 33
 physical structure, 48
Segregation, 13
Seismology model, 134
Slabs, pre-crustal, 82
Slag, surface, 84
Sheet, turbulent, 21–22
Signature, geological, 188
Size-mass relation, 2
Size, terrestrial planets, 148
Solar system, 1, 6–12
 budget, 6
 disc, 37
 disc, development, 44
 disc, early, 63–64
 disc, parameter variation, 41–42
 instability limit, 20
 ice-rock mix, 154
 onset, 22
Stars
 birthplaces, 10
 formation, 9
Stony bodies, 2
Stress, limiting, 5
Structure
 cumulate, 12
 during core formation, 74
 Jovian planets (now), 48–49
 physical sequence, 146–151
 radial variation now, Terrestrials, 147
Style, geological, 193

Sublayer, 165, 170
 base temperature, 165
 mechanism, 172
 thickness, 174, 178
Sub-planetary material, 30
Sun, 1–2, 4, 6–9, 38, 42–43, 76
Supernova, trigger, 10
Surface
 heat flux, 162
 temperature, Mars, 101
 zone, 82, 165

Temperature
 distribution, during core formation, 74
 fluctuations, 167
 global mean, 159
 Jovian discs, 46
 profile, upper mantle, 164
 profile schema, 162
 surface, 159
Temporal development, Jovians, 46
Terrestrial planets
 density structure now, 147
 molecular escape, 56
 physical model parameters, 142
Tethys, 2, 153
Tharsis region, Mars, 1
Thermal history
 identification, 178–188
 model, 158, 177
 simple model, 173–177
 terrestrial planets, 157
Time scale
 carbon dioxide sediments, Mars, 102
 core radius, 162, 189
 cratering, 34
 crustal freezing, 92
 deformation, 5
 Jovians development, 45
 nuclear, 8
 magma convection, 169, 189
 mass escape, 58, 60
 pre-crustal fractionation, 84
 particle collection, 33
 solar disc, 40
 thermal, 174, 189
Titan, 2, 12, 76, 81
 atmosphere, 97

Titan—*contd.*
  atmosphere–crust model, 104–105
  physical structure, 154
Transmissivity, atmosphere, 81, 87
Trapping
  matter, 21, 27, 31
Triton, 76
Turbulence, 12, 19–22

Ultramafic rocks, 109
Upper mantle
  fractionation ratios, 109
  sample, 80
Uranus
  chronology, 76
  density, 2
  depletion times, 65
  disc model, 45–46
  emergence, 22–30
  mass loss, 57
  molecular escape, 57
  moons, 97
  physical structure, 48

Velocity distribution (molecular), 200
Venus
  atmosphere, 81, 97
  atmosphere model, 99
  carbon dioxide amount, 88
  chronology, 76
  core mass, 75
  core onset, 73
  crustal rearrangement, 172
  density, 2
  density structure, 142, 147–150
  emergence, 70
  energy, 161, 188
  flushing, 61–62
  global information, 177
  heat flow, mean, 162
  magma convection, 169
  mantle recyclings, 191
  molecular escape, 57

Venus—*contd.*
  physical structure development, 146–150
  pool model, 127, 129
  style, 194
  thermal history, 176, 178–183
  thermal time scales, 189
  temperature, global, 164
  volcanism duration, 191
Vesta, 2, 31, 154
Virial theorem, 8, 39
Viscosity, 4, 171
Volatiles
  loss, 9
  properties, 97
Volcanism
  central, 194
  duration, 175, 178, 190
  early, 193
  Mars, 191
  sheet, 194
Volcanoes
  Hawaii, 5
  Mars shield, ii
Vorticity transfer, 26

Water
  amount, 88
  distribution, 90
  oceanic, 89–90
  on Mars, 103
  properties, 97
  substance, 14
  vapour, 87
Work, 39

Xenolith, 81
X-gas, 13

Z-component, 6
Z-gas, 13

Academic Press Geology Series

Mineral Deposits and Global Tectonic Settings—A. H. G. Mitchell and M. S. Garson—1981

Applied Environmental Geochemistry—I. Thornton (ed.)—1983

Geology and Radwaste—A. G. Milnes—1985

Mantle Metasomatism—M. A. Menzies and C. J. Hawkesworth (eds)—1987

The Structure of the Planets—J. W. Elder—1987